BREAKFAST OF BIODIVERSITY

BREAKFAST OF BIODIVERSITY

The Political Ecology of Rain Forest Destruction

second edition

JOHN VANDERMEER
and IVETTE PERFECTO

foreword by
VANDANA SHIVA

Food First Books
Oakland, California

This book is dedicated to Tina and Kiko, and to the poor farmers of the tropics.

Text and cover design by Stephen Hassett.
Cover art copyright Corel.

Food First Books
398 60th Street
Oakland, California 94618
www.foodfirst.org

ISBN 10 : 0-935028-96-X
ISBN 13 : 978-0-935028-96-6

Library of Congress Cataloging-in-Publication Data on file with publisher.

Food First Books are distributed by Perseus Distribution,
387 Park Avenue South, New York, NY 10016.
(800) 334-4499
Printed in Canada
10 9 8 7 6 5 4 3 2

CONTENTS

List of Illustrations vi

Foreword by Vandana Shiva vii

Preface ix

1: Slicing up the Rain Forest on Your Breakfast Cereal 1

2: The Rain Forest Is Neither Fragile nor Stable 16

3: Farming on Rain Forest Soils 35

4: The Political Economy of Agriculture in Rain Forest Areas 50

5: The Multiple Faces of Agriculture in the Modern World System 68

6: The Political Ecology of Logging and Related Activities 81

7: Globalization and the New Politics 93

8: Rain Forest Conservation: The Direct or Indirect Approach? 106

9: Biodiversity, Agriculture, and Rain Forests 138

10: Who Constructs the Rain Forest? 161

11: Past Causes, Future Models, Present Action 171

Notes 179

References 189

Index 197

About Food First 204

ILLUSTRATIONS

Figures

1.1 The Lowland Tropical Rain Forest of Central America.
1.2 Map of Nicaragua and Costa Rica.
2.1 The Distribution of Rain Forests in the World.
2.2 How Nutrients are Absorbed by Plants.
3.1 Hypothetical Illustration of Typical Distribution of Soils under a Tropical Rain Forest.
3.2 A Production Plan Based on the Underlying Soil Mosaic.
3.3 Slash and Burn Agriculture.
3.4 Functioning of the Chinampa System.
4.1 Banana Production in Central America. (2 photos)
4.2 Aerial View of United Fruit Company Banana Plantation in Costa Rica.
4.3 Banana Workers Prepare Metal Monorail to Receive Harvested Bananas.
5.1 Goods and Money Transfers in the Modern Agricultural System.
5.2 The World System.
6.1 Informal Uses of Rain Forest Wood. (3 photos)
6.2 A Recently Logged Area in Southern Costa Rica.
6.3 A Bulldozer Moves a Recently Felled Log.
6.4 Natural Disturbance Is a Recurring Event in Many Lowland Rain Forests. (2 photos)
8.1 Sketch Map of the Region Surrounding the Town of Puerto Viejo.
9.1 The Relationship Between Planned Biodiversity and Associated Biodiversity.
9.2 How Biodiversity Changes as a Function of the Intensification of the Agroecosystem.
9.3 Patchwork of Forest Fragments.
9.4 The Traditional Coffee Agroecosystem in Central America.
9.5 The Modern, Chemically Intensive Coffee Agroecosystem in Central America. (3 photos)
9.6 Number of Species Preserved within Protected Areas.
11.1 The Web of Causality for Deforestation of Rain Forests.

11.2 The Expanded Web of Causality for Deforestation of Rain Forests.
11.3 Poster on the Wall of CORBANA.
11.4 Poster on the Wall of the Ministry of Agriculture in Nicaragua.

Tables

6.1 The Approximate Damage to Biodiversity.
9.1 Insect Species Found in the Three Coffee Systems.
9.2 Hypothetical Biodiversity Preservation.

FOREWORD

by Vandana Shiva

John Vandermeer and Ivette Perfecto have in their book *Breakfast of Biodiversity* raised some of the most challenging questions about conservation of biodiversity.

The old conflict of ecology versus economics, environment versus development, has, in the post Rio era, mutated into a conflict about how natural resources will be conserved, by whom and for whom.

The dominant paradigm sees conservation as dependent on financial investments, which are, in turn, linked to increased economic growth, international trade, and consumption. This approach allows continued destruction of the environment and people's livelihoods in the domain of the productive economy, while allowing islands of "set-asides" and "wilderness" reserves, which also displace people and destroy indigenous cultures and lifestyles. In this paradigm, there is an artificial separation of conservation and production, of people and nature. Further, the increased trade and commerce that generate the financial resources central to this paradigm of conservation themselves destroy natural resources, biodiversity, cultural diversity, and people's livelihoods. The result is nonsustainable islands of biodiversity threatened with constant erosion by a sea of pollutants and monocultures.

The second paradigm is based on conserving biodiversity as the very basis of production, which ensures that both nature and people's livelihoods are protected. It is this second paradigm that is both people- and nature-friendly that this book articulates. In this paradigm, conservation cannot be isolated from production, investment, and trade. Environmental action becomes inseparable from issues of social justice and peoples' economic rights. As the authors state in closing, "calls for boycotts of tropical timbers or bananas need to be coupled with actions to change investment patterns and international banking pressures."

PREFACE

The world is already saturated with books about rain forests. What then can be our excuse for writing yet another? It's simple. After reading many of those other books in the course of preparing slide shows, lectures, and discussions, we came to believe that for the most part they really did not get it completely right. While they all made very important points about the nature of the rain forest and the alarming rate of its disappearance, and they all had particular analyses about causes, their analysis focused on one or another issue—overpopulation, export agriculture, peasant agriculture, etc.

We can appreciate the temptation to focus on the facts of rain forest destruction, and we agree that the nature of the problem itself is quite worthy of persistent propaganda. This, we suppose, is why all the books say the same thing—tropical rain forests are useful and beautiful, yet they are being destroyed. That the problem needs to be brought to the attention of the public, we agree. But once alerted to the problem, the public asks what to do. Causes must be addressed, and we feel that most of the popular literature on the subject does not do it adequately.

To be sure, many books talk about causes. But most frequently, authors are concerned with identifying some "ultimate factor"—overpopulation, greedy lumber companies, inefficient peasant farmers, avaricious export agriculture. We agree that some, even many, of these forces are part of the picture. But in the final analysis the cause is far more complicated. Indeed, the nature of the complications is the cause. This sort of analysis is multifaceted, with many interconnecting components—what we refer to as the "web of causality." However, we also feel that it is not difficult to appreciate this analysis, if the focus is not on an individual component but rather on the complete web. That is the purpose of this book.

We aim to alert the reader to several obvious facts. The web includes subjects that ordinarily do not occur together between the same book covers: the poor soils of peasant farmers, international diplomacy, international agricultural economics, and a variety of other strands in the web.

Thus our narrative ranges from the acidity of rain forest soils to the acridity of international politics.

Furthermore, this book is in no sense complete. It is not a comprehensive analysis of the nature of rain forests and their potential utility, nor is it a full presentation of the details of rain forest destruction the world over. Many other books do precisely that, and they do it quite well. Our purpose is to elaborate, in a compact and straightforward manner, the complicated story of why rain forests are disappearing. This is mainly a social, economic, and political story, with a pinch of ecology. The story has nothing to do with overpopulation and is not about a few evil capitalists who care more for profits than trees. Rain forests thus will not be saved by handing out condoms nor by refusing to buy furniture made of tropical wood. The only way to reverse the pattern of the past five hundred years is, first, to understand the complexity of the web that creates the problem in the first place and, second, to develop a strategy that shreds that causal web. Perhaps a boycott of tropical woods would make sense, but only within the context of a clear analysis of how that boycott contributes to eliminating the web of causality.

While our purpose is to elaborate the web of destruction in a brief form, we frequently find it easiest to argue from example. Since our personal experience is mainly in the tropical rain forests of Central America, almost all of our examples are from there. Other authors would obviously have used different examples. But in the end, our analysis is not restricted to the Central American case. To be sure, there will be differences from place to place — logging is more important in Southeast Asia, for example, cattle ranching more important in Brazil. But the general principles, and consequently the appropriate actions, are not dependent on the particular site. The web is strung slightly differently in each place, but its connecting strands are similar the world over.

It would be extremely difficult for us to remember the many individuals who have contributed to our formulation of this issue. For the most part they know who they are, and we thank them. In particular, we wish to thank the "Bluefields group" at the University of Michigan; the professors and students in the Department of Ecology and Natural Resources at the Universidad Centroamericana, Nicaragua; several farmers from El

Progreso, Costa Rica; the members of La Fonseca and La Union cooperatives near Bluefields, Nicaragua; and finally several Mexican farmers who alerted us to a more sophisticated vision of the aesthetics of rain forests than we had seen earlier. Special thanks are due to Victor Cartín, Marianos Azofefa, Ernesto Lemos, and Doug Boucher. Mike Jody patiently untangled much of our prose, for which the reader should be just as grateful as we are.

John Vandermeer and Ivette Perfecto
Ann Arbor, Michigan
February 1995
April 2005

1: SLICING UP THE RAIN FOREST ON YOUR BREAKFAST CEREAL

THE EASTERN SLOPES of the Barva volcano catch water-laden trade winds from the Caribbean to create the climate of Costa Rica's eastern coast, location of some of the most beautiful rain forests in the world. Here you can experience that special feeling that inspires poets and explorers—from the myriad vegetative forms so evident even at first glance (Figure 1.1) to the misty mornings that invoke mysterious feelings and bucolic images of paradise lost. The trade winds rise as they encounter the eastern seaboard, and with their ascent they cool, condensing the water vapor they borrowed during their voyage across the Caribbean. The consequent rains collect in several basins and come together roughly at the town of Puerto Viejo before continuing northward to empty into the San Juan River, the border with Nicaragua. This is the region known as the Sarapiquí (sah rah pee KEE), site of several of the world's most famous rain forest conservation areas (Figure 1.2).

Streaming into the area to partake of the breathtaking beauty of the natural world in this, "one of the last" pristine places in the world, are biologists, ecologists, and ecotourists, spending their grant money or retirement savings to visit the "heritage of humanity."

The rain forest here, as elsewhere, is a collective human construct that sometimes serves as our mystical Eden, but it is also a material collection

Figure 1.1. The lowland tropical rain forest of Central America. Note the wide variety of vegetative forms evident.

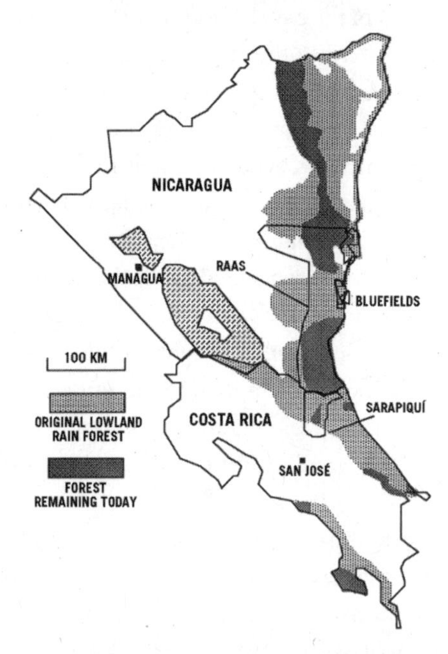

Figure 1.2. Map of Nicaragua and Costa Rica, illustrating the original and current extent of lowland tropical rain forest, and the position of the two political units, the Sarapiquí county of Costa Rica and the RAAS (Regiön Autonoma del Atlántico Sur) of Nicaragua (further discussed in Chapter 8).

of fabulous plants and animals—a natural outcome of the high temperature and heavy rainfall of equatorial climates. It is hardly necessary to repeat the cliches any longer—tropical rain forests cover only 7 percent of the earth's surface yet harbor at least 50 percent of the world's plant and animal species (the earth's biodiversity); they are the lungs of the world, eating away at the excessive carbon dioxide we have excreted from our industrial metabolism; they are the source of foods and pharmaceuticals, bananas and Brazil nuts, chocolate, cashews, coffee and cocaine, cortisone and quinine. They are also beautiful! The aesthetic loveliness of these forested lands cannot be overestimated, and the sense of wonder one experiences walking through this cradle of biodiversity cannot be expressed in words.

But, as anyone visiting the Sarapiquí can readily see, all is not well in this garden of Eden. Certainly, it remains beautiful inside of the conservation areas. The problem is outside those areas. And the problem is the same one Costa Rica has had ever since Minor C. Keith built his famous railroad and helped found the United Fruit Company in 1898. The problem is the banana. Large banana companies have converted vast areas to banana plantations, thus threatening both directly and indirectly the rain forests we so revere. Those same biologists, ecologists, and ecotourists who love the rain forest when they're in Costa Rica also love to slice bananas onto their cereal in the morning. And with our penchant for viewing the world in isolated little disconnected fragments, it is apparently difficult for us all to see the connection between the knife that slices the banana into our cereal bowl and the chain saw that slices tree trunks onto the rain forest floor.

Not so long ago, environmental activists in the developed world became aware of the so-called hamburger connection. Central American rain forests were being cut down at an alarming rate to make way for the production of low quality beef to supply the fast food industry in the first world. Stop eating fast food hamburgers, the argument went, and you would reduce the demand to cut down more forest. The banana expansion currently underway in Central America has been likened to this hamburger connection. But the whole argument surrounding the hamburger connection was flawed, and an attempt to construct the same argument for bananas would simply repeat those flaws. In fact, the expansion of bananas, like the expansion of pasture for beef production, involves a tangled web of subtle connections. Tweak the web at one point and it reverberates all over, sometimes in unexpected ways. Understanding the nature of the connections in this "web of causality" is the purpose of this book.

The transformation of the Sarapiquí is neither unprecedented nor unique, which makes it a useful example. Similar patterns occur throughout the tropics. Sometimes the pattern involves bananas, sometimes cattle, maybe citrus, African oil palm, or rubber trees—a variety of commodities similar politically if quite distinct biologically. The pattern is a six-stage process: (1) Visionary capitalists identify an economic opportunity for the market expansion of an agricultural product. In this case, the most recent opportunity was the opening up of markets in eastern Europe and the unification of western Europe. The product is bananas. (2) The capitalists purchase (or steal, or bribe their way into a government concession) some land, including land that may contain rain forest, which is promptly cut down. (3) They import workers to produce the product (in this case workers come from all over Costa Rica and even from Nicaragua). (4) After a period of boom the product goes bust on the world market, which means scaling back production, which in turn means releasing a significant fraction of the workforce. (5) The newly unemployed workforce seeks and fails to find employment elsewhere and must seek land to grow subsistence crops to tide themselves over until other work can be found. And finally, (6) the only place the now unemployed workers can find land no one will kick them off of is in the forest, which means yet more of the rain forest is converted to agriculture.

In this way Costa Rica, one of the world's showcases of conservation, has indirectly promoted a policy that actually encourages rain forest destruction. That is interesting by itself, but this specific example is not as important as the general idea it highlights. The crop in this instance happens to be bananas, but the general pattern is all too common.

Costa Ricans and Their Love/Hate of Bananas

An afternoon in Puerto Viejo, the little town located near the confluence of the rivers draining the Barva volcano, reveals something that might surprise a European or North American tourist. Despite the fact that, given their history, Costa Ricans understandably love to hate bananas, it was difficult to find anyone in town who did not fully support the massive banana expansion in the early 1990s. Furthermore, the government, both local and national, encouraged the expansion with a vigor normally associated with a depressed northern U.S. city courting a big assembly plant. *(You want no unions? You got it. You want tax breaks? Just say how much. You want license to pollute? Smoke your heart out. But please, just locate here.)* Costa Rica is as debt-laden as the rest of Latin America,[1] and needs all the money it can get just to service its debt. The expansion of bananas is one way to make money. Thus, despite the recognition that social and environmental problems will inevitably come along with the bananas, the vast majority of Costa Ricans, both in the Sarapiquí and elsewhere, welcomed the expansion. A small group of Costa Rican environmentalists protested, but they were overwhelmed by more powerful interests.

Of the approximately quarter of a million hectares[2] in the Sarapiquí valley, some 50,000 are devoted to biological preserves.[3] Another 100,000 hectares are in the legal hands of small peasant farming communities. The rest (approximately 100,000 hectares) is a mosaic of small farms, most without title to the land; secondary and old growth forest; cattle pasture; and an occasional sizable ornamental plant or fruit plantation.[4]

On the periphery of the valley lies an extensive banana plantation owned by the Standard Fruit Company, a major employer in the region for the past quarter century. The history of Standard Fruit provides an example of what might be expected on a larger scale in the near future. Tales are common of pesticide abuse, waste dumping into local waterways, deforestation, and the massive social problems normally associated with a frontier area. Best documented is the celebrated case in which Standard Fruit was accused of negligence in its use of DBCP (dibromochloropropane), a popular fungicide.[5] During the early 1970s thousands of banana workers were rendered sterile by this poison. In 1993 they filed a class action suit against Standard Fruit (and others) in a Texas court. The companies agreed, in 1997, to pay US$41.5 million to workers who could demonstrate they had been rendered sterile, but thus far they have used a variety of legal maneuvers to avoid any payments. This and other past records indicate that historically the banana companies have not accepted responsibility for the health and safety of their workers, the community, or the environment.

Perhaps even more ominous is another long-term pattern evidenced by the Standard Fruit operation in the area. Standard Fruit (known more frequently by its brand banana, Dole) employs workers who migrate to the Sarapiquí from other parts of Costa Rica, and increasingly from other countries, especially Nicaragua. These workers are retained as long as the market for bananas is sufficiently robust, but are let go when sales slacken. The laid-off workers are mainly rural people, former peasants drawn into rural wage labor. In past decades the ebb and flow of the banana business has created critical periods in which many workers were laid off and forced to fend for themselves. These layoffs were a natural product of the world economic system, due both to fluctuating banana prices, and to the very existence of a two-part economy—export bananas on one hand and worker/farmer on the other. There is little employment opportunity in the area, other than the banana companies, so when workers are laid off they must either migrate to the city, adding to the growing shantytowns, or turn to farming. In order to farm they have to find a homestead. Sometimes that small piece of land is in a rain forest. Other times it is a small corner of some large absentee landowner's cattle ranch, in which case, depending on complicated criteria, either the homesteading family is eventually forcibly evicted or the state agrarian reform institute adjudicates a "fair" purchase for the peasant family.

The past thirty or forty years have seen this arrangement persist, with rain forest cover in the region plummeting from almost 90 percent in 1950 to approximately 25 percent today. Only a small portion of the remaining 25 percent is not part of one of the four extensive biological reserves.

Loggers, Farmers, and Banana Companies: A Rich History[6]

This pattern, so readily observable today, is set in a rich ecological history beginning well before the current crisis.[7] Early in the century, extensive river systems were used to transport both logs and bananas. Logging was a rather small-scale operation by modern standards, but it had the effect of drawing workers to the area and creating pathways into the forest. Since only a handful of the many species of trees in the rain forest were actually valuable, it was necessary to scout out and then cut a path to the valuable trees and, after cutting them down, haul them out with teams of oxen or horses. By the 1930s, the land along almost all the rivers was deforested and planted with bananas, while the surrounding forest was riddled with trails made for dragging logs. The logging process intensified in the late 1940s and 1950s, when machinery was brought into the area and a complex network of logging roads crisscrossed what forests remained after the inroads already made by the banana plantations. Men who originally came to the area to work in the logging industry used these roads to gain access to logged areas and frequently established homesteads. Former banana workers did the same thing.

In the late 1940s everything changed in the Sarapiquí, as it did throughout the Atlantic coast of Costa Rica. Devastating fungal diseases routed the banana industry. The extensive plantations of the United Fruit Company and of various independent producers were decimated by this disease. No cure could be found, and the company moved its entire operation to the other side of the mountains, where the disease had not been established. It was not until the mid-1950s that a variety of banana that was genetically resistant to the disease was developed, thus enabling the Standard Fruit Company to reestablish its plantations in the area in the late 1950s.

The plot began thickening in the early 1990s. In anticipation of an expected surge in the demand for bananas (the anticipated result of opening markets in eastern Europe and the economic unification of western Europe), five or six major banana companies began purchasing large expanses of land and expanding banana production accordingly. The area planted to banana rose from 20,000 hectares in 1985 to 32,000 hectares in 1991, and visits to the area in 1991 and 1992 revealed intense activity in setting up new banana plantations throughout the valley. As much as 45,000 hectares were expected to be in bananas by the end of 1995.[8] According to a report from the Costa Rican banana research institute (CORBANA), a total of 46,557 hectares were actually in production in the Atlantic Coast region of Costa Rica by 1997.

As had happened in the past, workers were drawn from all over the country. But breaking with past traditions, this time there apparently were not enough Costa Rican workers to do the necessary work, and workers were also attracted from Nicaragua, Panama, and even Honduras. It appears that by now almost all of the arable land not currently in either biological preserves or organized peasant agricultural communities has become banana plantations.

A variety of factors make Costa Rica, and particularly the Sarapiquí basin, an especially attractive target of the banana companies. Notably, the infamous Solidarista movement has destroyed all union activity in the area. Some twenty years ago, this church-based, U.S.-supported, antiunion movement systematically moved into the Sarapiquí valley to replace all banana labor unions with a new concept for worker organization. Solidarista dogma outlaws strikes, does not recognize the right of workers to bargain collectively, and seeks to attract workers with frivolous benefits such as clubhouses and soccer fields. With massive funding from the Association for Free Labor Development, which is the international wing of the AFL-CIO and has long been suspected of having CIA ties, democratic labor unions were systematically attacked throughout Costa Rica. The campaign was especially effective in the Sarapiquí, where union membership now stands virtually at zero and company officials proudly proclaim that no union people are able to find jobs.[9] A local political official told us in 1991 that the planned banana expansion would have been impossible without the existence of the Solidarista organizations.

A second important factor was the willingness of the Costa Rican government and its partner, the United States of America, to create infrastructural conditions that favor the banana companies. Roads were constructed, bridges were built, hospitals and schools were planned, all for the purpose of creating an attractive infrastructure for the banana companies. The U.S. Army Corps of Engineers was enlisted in this effort. In a 1992 program called "Bridges for Peace," army engineers built roads and bridges in the Sarapiquí. A cynical U.S. serviceman told us the program has been unofficially dubbed "Bridges for Bananas," as the construction so obviously focuses on improving infrastructure for the export of bananas. U.S. Army engineers built many of the roads and bridges that originally carried the logs of the cut rain forest, and now carry the harvest of the international banana companies. Indeed, with the infrastructure provided by U.S.

taxpayers at the request of the Costa Rican government, from roads and bridges to the Solidaristas, from the "converted" rain forest to new social infrastructure, investment opportunities look good—that is, if you are a banana company.

But the banana companies, mindful that their operations might attract outside attention, were prepared to pay "expert" scientists to mollify the public. Corporación Bananera Nacional (CORBANA) was formed in 1990. Some twenty years earlier a smaller national effort, Asociación Bananera Nacional (ASBANA), had been formed by the Costa Rican government for the purpose of developing technical assistance for small producers of bananas in the country. Operating on a tiny budget, this small research operation persisted until 1993, when the rush to privatization caught up with it and ASBANA changed its name to CORBANA and began to receive money directly from the banana companies. For every box of bananas exported, each company pays a fee to support the research efforts of CORBANA. Theoretically, CORBANA conducts research aimed at making banana production more environmentally friendly. This research was to include proposed projects on using biotechnology to develop strains of bananas resistant to pests, development of organic fertilizers, and extensive surveys of fauna in the banana plantations (ostensibly to monitor the effects of the plantation on wildlife). However, in a visit to the CORBANA facility we observed very impressive projects on soils, plant diseases, parasites, and drainage, but none of the celebrated studies to promote environmental friendliness.

But we repeat and emphasize that the expansion of bananas has always been viewed as a positive event by nearly everyone in the Sarapiquí and by most observers in the entire country. Local workers and peasants see jobs being created, local merchants see a potential surge in business, and local politicians see an increase in their power base. The Costa Rican government itself promotes the expansion since it sees the increased tax revenues as helping to pay debt service on its tremendous international debt. The accepted fact that almost all profit from banana farming will leave the country seems of little local concern. This is perhaps understandable given the state of the local and national economy. But less comprehensible is that segments of the international "conservation" community have come on board and either retain a calculated obliviousness to what is going on or actively pursue a neutral position. Significant yet weak opposition is coming from a small, loosely structured local conservation movement composed of Costa Ricans and organized without the help of the international conservation community. They are fighting against what appear to be insurmountable odds.

This example illustrates the dynamics that occur, with different details of course, throughout Central America and in much of the rest of the world. Because of the nature of the world economic system, Costa Rica really has no choice but to promote the expansion of bananas. Costa Rica's international debt, accumulated because of its position in the world economy, and its need for the expansion of international capital, require that it seek tax revenues however it can. The banana companies themselves (at least one of which is Costa Rican) continue to play their historical role as international accumulators of capital and temporary employers of peasants, thus maintaining the dysfunctional two-part economy. Peasants continue arriving from other parts of the country and even from other countries, seeking jobs and the "good life" and willing to accept minimal conditions; but since the Solidarista movement destroyed the unions, they are without significant political representation. The stage was initially set by the loggers with their systems of logging roads, and the first wave of banana plantations with their periodic layoffs, which forced peasants into the forests. If the process continues with this basic overall structure—and we see no reason to doubt it will—there is little hope in the long run even for those rain forests under protected status.

Despite what some promoters claim, banana plantations do not last forever. A variety of ecological forces eventually catch up with such intensive production, and the plantations must be abandoned. Will the legacy of bananas leave the area with much degraded conditions of production, as has happened repeatedly in the past? Who will bear the costs of restoring ecosystems to their original state, if that will even be possible? And who will share the perspective of the thousands of rural people deprived of their land and their livelihood, with no place to go? Who will tell them that the rain forest is more important than feeding their families?

The Problem from Various Perspectives

In the midst of these dramatic events, an internationally recognized ecologist gave a public lecture at a local ecotourism center in the Sarapiquí, claiming that recent deforestation in the area was due to the inevitable march of Malthusian reality. He claimed that overpopulation was causing the destruction of the forest. In a trivial sense, of course, such an observation is true. There undoubtedly is an overpopulation of banana companies, an overpopulation of former banana workers looking for land, an overpopulation of adventurers seeking their fortunes in a new frontier zone, an overpopulation of greedy people and institutions, and even an overpopulation of ecotourists from Europe and the United States.

But when this expert ecologist declared that a Malthusian crunch was the root of the problem, he was actually implying something rather different: that the pressure of having too many children—the birth rate of the population—is the real problem. This point of view implies that the main solution to the problem is birth control. It further implies that this is a sufficient solution, that it is useless to do anything other than promote birth control, and that as long as population densities remain as they are, the pattern of deforestation will continue.

An alternative viewpoint expressed to us by a local conservationist is that avaricious banana companies are cutting down rain forests because they are hungry for profits. They will stop at nothing to satisfy their need to accumulate ever greater quantities of capital, and the forests will continue to disappear as long as the banana companies are allowed to continue their greedy operations. This also is a distinct point of view. It implies that the only solution to the problem is to eliminate the capitalist. It further implies that this is a sufficient solution, that it is useless to do anything other than "smash capitalism," and that as long as the drive to accumulate capital remains, the pattern of deforestation will continue.

These points of view are prisms through which the facts of the matter may be interpreted. They both encourage a single-focus solution—stop population growth, or smash capitalism. We believe that both are right in a very limited sense. But we also believe that they are both wrong in a broader and more practical sense. Ultimately each of these prisms focuses on a single thread in a fabric of causality. Eliminating one thread will not eliminate the problem. The problem is the fabric itself. The proper means of understanding the situation, then, is to look at the complicated way that various forces are interdependent, especially focusing on the way countervailing tendencies are resolved. The approach we take in this book may at first glance seem as narrow as the approaches of those who advocate population reduction or smashing capitalism. We assert that food insecurity and poverty are the root causes of deforestation. It is a critical thread in the fabric of causality.

We take this approach for two reasons. First, we wish to provide an antidote to the simplistic views that either overpopulation or avaricious capitalism cause deforestation. Second, we will argue that, given the ultimate goal of reworking the entire fabric of causality, the place to begin that process is with food security. We will not argue that assuring food security for the peasant class will per se stop deforestation, but rather that beginning the political process of reorganizing socio-economic-ecological systems by examining questions of food security will force both analysis and practice into the realms ultimately necessary for the resolution of this issue. When neo-Malthusians suggest there are too many people for the land base, the food security position reveals several important particulars: that peasants seek land to feed their families not because there are too many of them or too little land (at least right now), but because available land is occupied by other activities. Our orientation will also reveal that the techniques for sustainable agriculture in that zone have been replaced with destructive, chemically based ones, and further that the legal status of most peasants is "landless" even when they clearly occupy a piece of land. When radicals purport that avaricious capitalism causes deforestation, the food security position shows that the evolution of modern agriculture has created international structures that force even progressive governments like Cuba's to invite those greedy capitalists into their economies. The international order that causes food insecurity in the developing world is implicated in a chain of events that ultimately leads to the transformation of workers to peasants who must seek out rain forest land to farm in order to provide food for their families.

We do not wish to leave the impression, however, that food insecurity is just another mechanistic cause to which the problem of rain forest destruction may be reduced. It is clearly not. But, as a mode of analysis, examining food insecurity will cause us to deal with the entire complex web of ecological, sociological, economic, and political issues on which the poisonous spider of rain forest destruction crouches.

Two Models for Saving the Rain Forests

Current events in the Sarapiquí region are, alas, not unique. Tropical rain forest areas around the globe are experiencing similarly complex socioeconomic forces, which threaten to continue or even accelerate the destruction of this most diverse of all ecosystems. In all of these areas there has been some reaction from local and international concerns. Unfortunately, much of this response is misdirected because it is based on a distorted image of the facts, and on an implicit ideology—what we call the *mainstream environmental movement approach*—which allows only a narrow range of possible courses. We feel there is an alternative philosophical approach, the *political ecology strategy,* which emphasizes basic issues of security: security of land ownership, and the consequent ability to produce food for local consumption.

The mainstream environmental movement has raised large sums of money to purchase and protect islands of rain forest, with little concern for what happens between those islands, either to the natural world or to the social world of the people who live there. We doubt this strategy has much chance of succeeding. It is likely that in the short term the landscape will be converted into isolated islands of tropical rain forest surrounded by a sea of pesticide-drenched modern agriculture, underpaid rural workers, and masses of landless peasants looking for some way to support their families. The long-term prospects, however, are worse, as the example of the Sarapiquí suggests.

Our alternative, the political ecology strategy, emphasizes the land and people *between* the islands of protected forest. We feel it has greater credibility because of its willingness to search out the interconnections in this complicated system. This point of view has been variously known as "ecological development," "sustainable development," or "eco-development," though all of these terms have been cynically adopted by even the most environmentally destructive agencies. Whatever we call it, this approach views the problem as properly a landscape problem, with forests, forestry, agroforestry, and agriculture as interrelated land use systems, and seeks to develop those land use systems so that conditions of production according to the needs of the local population may be maintained. The political ecology strategy challenges nonsustainable development projects, such as modern banana plantations, and seeks to organize people to oppose ecologically and socially damaging development.

These two points of view lead to quite different projections of what the future rain forest areas of Central America might look like. If the mainstream position remains dominant, we expect to see, in the short term, a sea of devastation with islands of "pristine"[10] rain forest, and in the long term nothing but the sea of devastation. The political ecology point of view envisions a mosaic of land use patterns: some protected natural forest, some extractive reserve, some sustainable timber harvest, some agroforestry, some sustainable agriculture, and, of course, human settlements. This mosaic would be sustainable over time.

But is this a practical vision in the real world? The decision to promote bananas in the Sarapiquí can hardly be faulted on "modern" economic grounds. Sadly, however, if national and international commitment to the archaic economics of Adam Smith and the International Monetary Fund (IMF) persists, we fear continuing destruction of the rain forests and the deterioration of the lives of the people of Sarapiquí. The alternative requires a radical rethinking of what sorts of economic and political arrangements are to be tolerated, the sort of rethinking that can get you in trouble in Central America, the sort of rethinking that may even challenge the idea that it is our inalienable right to slice bananas onto our breakfast cereal.

Costa Rica, Bananas, and a General Pattern

The case study elaborated in this chapter is typical. Granted, there are cases in which rain forests are being cut with a profoundly different logic (several areas in Southeast Asia and much of the Amazon for example), but both historical and contemporary patterns the world over reflect the basic paradigms seen in this example. The details vary, but the underlying logic is remarkably consistent.

Costa Rica has been held up as one of the world's best examples of rain forest conservation. Its internationally recognized conservation ethic, its position of relative affluence, its democratic traditions, the remarkable importance of ecotourism to its national economy, and its willingness to adopt virtually any and all programs of conservation promoted by western experts make it the most likely place for the success of the traditional model of rain forest conservation. The fact that the model has been an utter failure in Costa Rica, where it had the greatest chance of success, calls the model itself into serious question.

This case study is intended as an outline of the general problem. In the rest of the book we discuss the details of how the problem comes about, ecologically, economically, and politically. Hopefully, at the end, what must be done to protect rain forests will be clear. Hopefully, the reader will come to agree with us that purchasing tracts of land and putting them under armed guard is folly. Hopefully, it will become apparent that stopping individual logging companies and avaricious agroexporters can be only a small part of the solution, and that basic questions of land and food security are the most central components of any potentially effective political strategy. Hopefully, it will be apparent that such political strategies begin to look more like past political strategies that helped stop the war in Vietnam, curtailed U.S. intervention in El Salvador and Nicaragua, and currently challenge the international hegemony of institutions like the World Bank, IMF, and World Trade Organization (WTO). And hopefully, it will be clear that only by our uniting with political forces that have similar fundamental goals can the future of the world's rain forests be brightened.

2: THE RAIN FOREST IS NEITHER FRAGILE NOR STABLE

THE TROPICS COMPRISE the area between the Tropic of Cancer and the Tropic of Capricorn. The lowland humid tropics, site of the world's rain forests, account for less than one-third of tropical lands—deserts, savannahs, and mountains also occur in the tropics. Tropical rain forests are evergreen or partially evergreen forests in areas that receive no less than 250 centimeters a year of rainfall, and have a mean annual temperature of more than 24 degrees centigrade with no frost.[1] Tropical rainforests are found in three general regions of the world, as shown in Figure 2.1.

Rain forests are enormously complicated creatures. Neither the people who live in them nor the scientists who study them understand everything about how rain forests work, which fact undoubtedly contributes to our mystical feelings about them. Yet comprehension need not elude us completely. Few of us understand how our automobiles work, but we do know certain basics—the motor will start when we turn the ignition, and the accelerator pedal is quite distinct from the brake pedal. In this sense we can also comprehend the way rain forests function. No one knows all the details, and all is not yet understood, but what is known with some certainty can be explained about as easily as explaining how to put the key in the ignition, or when to step on the brake rather than the accelerator.

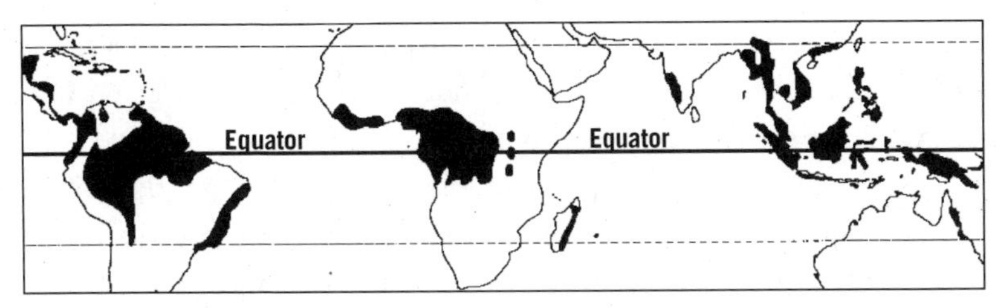

Figure 2.1. The distribution of rain forests in the world.

Most popular accounts of rain forest loss and/or preservation emphasize either a romantic or a utilitarian notion of rain forests, concentrating mostly on the larger political forces driving their destruction. This has led to some confusion and no little dogma about what can and cannot, or what should and should not, be done to save or restore rain forests.

Stable Web or Fragile House of Cards?

Since ecology began to fire the popular imagination in the 1960s, two major perceptions about the nature of ecosystems have resurfaced with remarkable regularity. These two ideas are in fundamental opposition to one another, yet this conflict frequently escapes notice. On one hand, ecosystems are thought to be fragile and easily damaged by the careless hand of man or woman. On the other hand, ecosystems are considered stable organisms, honed by evolution to be harmonious: the "balance of nature" is considered a stable equilibrium. If an ecosystem is stable, it is obviously not likely to be fragile. Yet these conflicting notions surface repeatedly in the popular ecology movement.

Perhaps nowhere are these two ideas applied with such misguided fervor as in the popular literature on tropical rain forests. Curiously, both the fragility and the stability of rain forests are attributed to their characteristic high biodiversity. Two common metaphors may help explain the rationale behind this idea. On one hand, the immense biodiversity in a rain forest may be thought of as the filaments in a spider web: the more connections, the stronger the web. Thus, because the rain forest has so many strands and so many connections, it must be very stable. On the other hand, the immense biodiversity may be represented as a house of cards, each card balanced precariously on edge, all the cards supporting one another to keep the house standing. This house of cards may be balanced because of the large number of cards, but removing just one could very well cause the entire structure to come tumbling down.

So which metaphor makes the most sense in fact? Is the rain forest a highly connected web that gains its strength and stability from its multiplicity of connections? Or is the rain forest a house of cards, precariously balanced and subject at any moment to total collapse if a single card is removed?

It may surprise anyone who has read a few books about rain forests to learn that these popular visions are both probably way off the mark. In the past twenty years ecologists have learned a great deal about how rain forests work. Added to what has already been gleaned from years of tropical forestry and from hundreds of years of knowledge of the indigenous people who live in these ecosystems, we now have a relatively clear understanding of the basic functioning of tropical rain forests. To be sure, an enormous number of particulars remain enigmatic—something to be expected from the sheer quantity of elements involved. But the overall picture is probably better understood than might be surmised from a look at the rain forest picture book on Uncle Ed's coffee table. Many of us do not understand how a carburetor works, yet we can easily grasp the function of the accelerator pedal. Similarly, many details of rain forest function are still poorly understood, yet we can fully grasp the nuances of biodiversity and other general features of rain forest function, and perhaps we can use that knowledge to our benefit.

Six Key Factors in Rain Forest Function

The overall picture of rain forest function can be understood in terms of six key factors. First, as is commonly mentioned in popular accounts, rain forests exhibit an enormous amount of diversity—different kinds of plants and animals and other living things. This diversity in and of itself generates some problems for individual species. For example, if hundreds of distinct species are to exist in the same place, that means most of them must be rare. If there is room for just 200 trees in a plot of land, and you wish to fit in 100 species, each species, on average, can only be represented by two individuals. This brings up the second factor: sex. How do such rare individuals find mates? The dominant elements of the forest are plants, and especially trees, and sex in plants happens through the process of pollination.

The third key factor is the problem (from the point of view of plants) of herbivores—the many insects and animals that enjoy eating leaves, shoots, and seeds. How plants deal with herbivores is a major piece of the story of how rain forests work. Herbivores tend to specialize on particular species of plants. For example, the mahogany moth eats nothing but mahogany trees. If mahogany trees are interspersed with numerous different species, as is the case when most species are rare, mahogany moths will have trouble finding their dinner. If mahogany trees are all clumped together, they make an excellent target for herbivores. This leads to the fourth factor: the dispersal of offspring. If seeds just drop to the base of a plant, and the young plants grow there, any insect that happens to find the adult will find the offspring too. So plants have evolved various strategies to disperse their seeds.

The fifth factor is death. In the context of the forest, a dead tree signifies much more than simply the passing away of one individual in a population. A dead tree means a gap in the canopy of the forest. Sunlight may now enter that gap, and for a time, the understory of the forest is quite different than it was before the death of the tree. Thus, a major element of rain forest function is the formation of so-called light gaps by falling trees. What happens in these gaps is of utmost importance to the structure of the forest.

The sixth key factor in rain forest function is the soil. The matrix from which the trees must draw their sustenance is very special in tropical rain forests. The same physical forces that make the rain forest itself so incredibly lush also make the soils on which they live rather infertile—an apparent contradiction that we explain below.

So these six factors—high biodiversity, pollination, herbivory, seed dispersal, light gap dynamics, and soils—are the simple elements shaping how tropical rain forests function. We note in passing that each of these six key elements will play a part in any and all proposed management schemes of tropical rain forest areas. They will be useful in understanding the enigma of poor farming potential in the most lush ecosystem on earth, the headaches of reforestation after commercial logging, and how to make good on the promise of food from the forest.

Factor 1: Biodiversity

We begin by examining the most obvious factor. In a forested area in northern Michigan, we found eight species of trees in a sample of one hectare (approximately 2.47 acres). In the same size area in the rain forest of Nicaragua, we encountered about 200 species of trees.[2] Entomologists netting insects in Kansas found ninety species of insects, while when applying their nets in the same fashion in a rain forest in Costa Rica they found 545 species.[3] In a Peruvian rain forest entomologist E. O. Wilson identified forty-three species of ants in a single tree, almost as many as those reported for all the British Isles.[4] In the Americas, bird species increase approximately fivefold from the middle latitudes (e.g., Kansas, Michigan) to the tropics.[5] While there are exceptions to this rule (for example, the number of small rodents running around in the leaf litter of a tropical rain forest is about the same as, or even lower than, the number running around in a northern forest), this general pattern is repeated for most kinds of organisms one examines—trees, herbs, insects, birds, etc.[6]

Ecologists have been debating the significance of the high degree of rain forest biodiversity for a long time, and so far have not come up with a satisfying explanation. What causes this great amount of diversity is a persistent and still enigmatic question. Initially it seemed that the high productivity afforded by large quantities of rainfall and hot temperatures would promote much diversity. However, when ecologists examined this idea closely it did not prove to be true. Indeed there are convincing arguments that greater productivity could actually result in much lower diversity. From the point of view of how much material is actually produced per unit time, the most productive ecosystem in the world is a modern cornfield. Other ecologists have suggested that the longer duration the tropics have been free of the ice sheets that covered much of the temperate zone during the great ice ages has allowed for the evolution of more different kinds of organisms. But here too there are opposing arguments, and this "time hypothesis" remains at best controversial. We could go on with a variety of other explanations, but the story is the same for each one of them. While Newton was able to explain the laws of gravity, and Darwin explained how biological organisms evolved, no comparable explanation yet exists for why some places (like the tropics) have so many species, while other places (like the arctic) have so few.

Though the origin of tropical diversity remains an enigma, the answer to the related question of how all these species are maintained in the same system is slowly being worked out. As mentioned above, the assumption that the more things there are in a system, the more stable the system has been accepted in the popular literature on rain forests for years. But a careful analysis of how ecosystems work has led us to the surprising conclusion that the fundamental theories of ecology actually predict the opposite. In fact, the more diverse the system, the more likely it is to be fragile. Our house of cards metaphor seems more likely to be true than the spider web.

Realizing that a highly biodiverse system like a rain forest is more likely to be fragile than stable, ecologists in the 1980s began trying to figure out how such diverse systems could be maintained. That is, rather than assuming that such complex systems were stable as a spider web and asking where they came from, we began admitting that maybe rain forests weren't so stable and asking how they could persist in the real world if they were house-of-cards fragile.

This is where ecology stands today regarding the issue of rain forest stability or fragility. The answers are certainly not yet well established, but the questions are clearer now than they have been before. Three research agendas appear to be dominating this issue today. First, many ecologists are asking questions about the particular ways in which ecosystems are held together—the details of the balancing cards, so to speak—for example, whether a tree with many rare herbivore species can coexist in the neighborhood of a tree with a few common ones, or whether a beetle that eats a bug that eats an herb is more likely to persist in the long run than a beetle that eats a beetle that eats the bug that eats the herb.[7] Second, other ecologists are focusing on the role of disturbance events such as storms and landslides, or even the inevitable tree fall–caused light gaps.[8] Since such disturbance events are as much a part of nature as the biological organisms that make up the ecosystem, it seems important to understand what their ultimate effect might be. Third, many ecologists have recently begun focusing on how the various species of trees in the forest are distributed in space. It turns out that many species occur in slightly aggregated clumps in the forest, and this pattern of distribution may help explain how so many of them can live together. It is relatively difficult for a particular kind of tree to make sure that its seeds are dispersed to all areas of the forest if most of the source trees for the seeds are located in a small clump. So, any given point in the forest is unlikely to receive seeds from all the species in the forest. Thus, the huge species diversity of the forest is not really so huge from the point of view of a single spot in the forest, and two species that are antagonistic to one another may only rarely exist in the same spot.[9] Furthermore, this tendency to form clumps of individual trees in the same species may increase the likelihood that some "natural enemy" will become devastating within that clump, thus opening up space for some other species to disperse into the area. Just as people concentrated in cities may be more susceptible to epidemic diseases, trees in a clump could also be attacked by a disease or devastating herbivore. This so-called enemies hypothesis could help explain how so many tree species are able to live together.

Factor 2: Pollination

Since the evolution of flowering plants over a hundred million years ago, terrestrial ecosystems have been dominated by a fundamental *mutualism*—animals help plants have sex and the plants reward the animals with food. The hummingbird drinks nectar from the flower of the wild banana plant and takes the pollen from one banana plant to another. The humming bird gets fed, and the banana plant has sex. Both species benefit from this interaction, the definition of mutualism. While the vast majority of plants in the world are pollinated by insects, the rain forest also contains special pollinating systems involving birds and bats.

In a rain forest, the problem of pollination is basically one of being rare and wanting to have sex (i.e., wanting to reproduce), and being unable to move. The problem is solved by a variety of means, but most commonly in one of four specific ways. First, particular species may be rare over a large range, but common in a local area. This leaves certain technical genetic problems unresolved and creates problems with elevated herbivory, but solves the simple problem of reproduction. Pollen can reach ova quite easily if a rare species always exists in a clump of its own.

Second, many rare species have evolved a process called *selfing.* Using a variety of mechanisms, many tree species are effectively hermaphrodites. This means that a single individual has the reproductive organs of both sexes. While this seems an obvious solution to the problem of reproduction, it creates enormous alternative complications associated with inbreeding. We all know what happens when cousins marry, and the genetic problems faced by the royal families who inbred are legion. Such problems are magnified enormously when an individual mates with itself. How plants deal with these new genetic problems is quite complicated and well beyond the scope of this introduction.

Third, many pollinating systems for rare plant species include synchronous mass flowering and a long-distance pollinator. For example some tree species typically display all their flowers in unison once per year, and all the individuals in the population of that species do so simultaneously. These trees typically use, as pollinators, bees capable of flying very long distances between meals. Thus, while the trees themselves occur only occasionally amongst a mass of other species over a wide area, when it is time for sex, all the individuals in the extended population put out all their flowers at the same time. The bees are able to fly from one tree to another, picking them out amidst the massed greenery of all the other species because of the abundance of colorful flowers in their crowns.

The synchronous mass flowering strategy has certain prerequisites. The pollinating bees must be supplied with a food source for the entire year. A particular species of tree will likely only flower for a week or two during that year. In order to have at least one species of tree flowering at all times during the year (an absolute necessity for the bee), there must be a minimum number of tree species in the forest. So, for example, if each of the tree species flowers for two weeks during the year, an absolute minimum of twenty-six tree species will be needed to maintain this particular form of pollination (assuming any overlap in flowering time is minimal). This requirement has dire consequences for any strategy that attempts to preserve small patches of forest. If, by accident, the patch is missing or loses just one of the necessary twenty-six species of trees, the whole system could collapse, since the bees would die during that two week period without food, and without their pollinators the trees would be unable to reproduce.

The fourth solution to the problem of reproduction for rare species is the habit of *traplining*. Some birds and insects are capable of creating a mental map in order to remember the locations of particular food sources. An individual plant that is currently producing flowers may be very far from others of its species, but a hummingbird, for example, may know exactly where the next individual of that species is located. The hummingbird thus has a number of individual plants in mind when she sets off in the morning to find food. She tanks up on nectar at the first site and then flies off to the next plant, remembering from day to day where these particular individual plants are located. She has, metaphorically, a trapline (trappers frequently refer to the series of traps they put out each night as a "trapline") formed by the flowering plants she visits each day. The individual plants, in turn, produce only a few flowers each day, attracting the pollinator but not monopolizing its attention, so that it may fly off to the next individual in the trapline.

These then are the four principal ways in which the problem of sex under conditions of rarity has been solved: local clumping, selfing, synchronous mass flowering, and traplining. Of course the details of pollination systems are far more complicated than we indicate here. Nevertheless, the basic patterns are really quite simple.

Factor 3: Herbivory

Herbivory as a way of life has existed for hundreds of millions of years. The spectacular elephants and lithe gazelles of the African savannas, and the cattle herds that graze America's western plains are well-known examples of herbivores. While insects are the most important pollinators in tropical rain forests, they are not universally beneficial to plants. As herbivores, insects are probably the plant's biggest threat to life.

Of the many important consequences of herbivory, one that has significance for humans is the evolutionary response of both plant and herbivore. Plants have evolved an overwhelming array of defenses against the many herbivores that potentially attack them. Three major types of defense have evolved: structural, chemical, and mutualistic.

Structural defenses include all of the spines and hairs that one knows so well from spiny vines, stinging nettles, and cacti. Such structures need no further explanation. But at a more microscopic level, small hairs on the surface of leaves, extra thick cuticle on the stem, and a host of other small-scale equivalents of spines have evolved as a protection against insect herbivores. What seems like fuzz to us is an array of stilettos to a tiny insect.

Only recently has the importance of chemical defenses been fully appreciated by ecologists. While we have known for years that many plants are poisonous, we did not realize that most of those poisons had evolved mainly as a defense against herbivores. Indeed, we now appreciate that most plants, especially those living in tropical rain forests, produce chemicals to protect against the ravages of herbivores—natural insecticides, so to speak. Indeed, some of our most famous chemicals, like caffeine and nicotine, are produced by plants mainly as a response to natural selection pressure from herbivores.

There is, however, a secondary consequence of this chemical defense. Herbivores themselves evolve. And just as one can often find a chemical antidote for a poison, herbivores have frequently been able to evolve antidotes for the poisons plants produce to protect themselves. This process is so common that ecologists have dubbed it the "coevolutionary arms race." Every time an herbivore evolves a detoxifying mechanism to deal with a plant poison, the plant is subjected to pressure to evolve a new poison, which puts pressure on the herbivore to develop a new detoxifying mechanism. This seems to be a continuously operative process in nature, and has significant consequences for those seeking to engage in agriculture in rain forest environments.

The third form of herbivore defense is mutualism. In a wide variety of cases, plants have evolved the capacity to use other organisms to protect themselves from herbivores. The most celebrated case is that of the ant and the acacia plant. Acacia plants produce hollow spines in which stinging ants live. Further, the acacias provide the ants with a self-contained energy and protein source (the acacia has glands that produce food exclusively for the use of the ants). In return, the ants act as a kind of armed guard for the acacias. They will viciously swarm over and kill any herbivore that makes the mistake of trying to eat the leaves of their host. This phenomenon is quite common and represents the plant taking advantage of the overall structure of the ecosystem. Every herbivore has natural enemies—parasites and predators that eat it. Plants may evolve structures to attract and co-opt these natural enemies, thus creating a mutualistic relationship between the plant and the predators of its natural enemy, a natural embodiment of the adage my enemy's enemy is my friend.

Factor 4: Seed Dispersal

While the problems related to seed dispersal exist all over the world, they are especially important in a tropical forest. Seeds are usually the most energy-rich part of a plant and are thus a prime target for herbivores. If seeds are simply dropped in a bunch at the base of the parent tree, they make an excellent target for the herbivorous insects and mammals that eat them. Consequently, most plant species have evolved mechanisms to disperse their seeds away from the fixed position of the parent plant. In tropical rain forests these mechanisms frequently involve animals, especially mammals and birds.

There is something of a contradiction involved in seed dispersal. While it is generally a good, even necessary, process, seed dispersers also have the potential of being seed predators. Squirrels eating acorns, for example, may very well eat all the acorns produced by a tree, leaving none to produce the future generation of oak trees. To solve this problem, plants have evolved two general methods: separation of seed from the disperser's food reward, and satiation. In the vast majority of cases, plants have evolved attractant structures to entice dispersers to scatter their seeds. Almost all of the fruits we eat are examples of this strategy. A bird that eats a berry digests the pulp of the fruit and passes the seeds in its stool. When a horse eats an apple, the lure of the apple is not the tiny brown seeds within, but the juicy pulp—the horse passes the seeds unharmed. The reward for the animal disperser has been separated from the seed itself.

In a significant number of cases the seed is the attractive structure—for example, oaks and their acorns. Since the disperser, in this case the squirrel, is by definition also a seed predator, a contradiction emerges: how to provide the attraction, yet get the seeds dispersed. This is actually a rather complicated issue, and its resolution is only partly understood. It involves a concept called *satiation.* Oak trees tend to have *mast years,* in which an unusually large number of acorns are produced by all the trees in a population. After three or four years of relatively low production of acorns, in a mast year all the trees in a river basin produce an excessive number of acorns. The consequences are not surprising. After a short supply of acorns for three or four years, the squirrel population has thinned out, and all the squirrels in the river basin are simply not capable of eating the large number of acorns produced. They get satiated. In a mast year the animals are satiated as seed predators and are thus able to act as seed dispersers. In the example at hand, the squirrels continue to gather and bury acorns, but have little motivation to go back, dig them up, and eat them. Thus they leave a significant number of acorns to germinate and produce new oak trees. This process of seed predator satiation is sometimes very important in rain forests, and is almost universal among the dominant trees of the Southeast Asian rain forests.

Factor 5: Gap Dynamics and Other Forms of Disturbance

Old forests typically have old trees, and trees do not live forever. A sudden gust of wind will bring an old tree down, creating a hole in the forest canopy, and bathing the understory with light. What follows is that sun-loving plant species, known as pioneers, enter the gap first, followed by secondary species, and eventually a "climax" tree grows up to take the place of the tree that fell. *Climax* refers to the ultimate stage in this successional sequence. Specific kinds of species are associated with each stage in the sequence, the climax species being the ones that are replaced by their same kind when they die. The process of replacing a fallen tree with a new adult tree takes anywhere from thirty to hundreds of years, depending on the type of forest. The continual process of a tree growing, becoming large, falling to create a gap, and eventually being replaced by another is thought to be a major force influencing how rain forests are structured, and therefore how they function.[10]

Recent studies suggest that over the last few decades, the turnover rate of these gaps has increased in tropical rain forests.[11] In other words, the gaps form more frequently than in the past. Although the reasons for this increase are not known, it has been suggested that the consequences will be a change in forest structure, and a possible change in function, with an increased predominance of light-demanding plants and the eventual extinction of slow-growing, shade-tolerant species. Plants absorb carbon dioxide and release oxygen. The carbon in the carbon dioxide is incorporated into the tissue of the plants. Trees with harder wood tend to sequester, or absorb and hold, more carbon than trees with softer wood. Light-demanding and fast-growing trees generally have softer wood than the shade-tolerant, slower-growing ones. Thus, if the structure of the forest changes from one dominated by slow-growing species to fast-growing ones, the forest will change the rate at which it absorbs carbon dioxide from the air. Since carbon dioxide is one of the main greenhouse gases, this change in forest structure may contribute to the increase in global warming.

In the past few years a great deal of attention has been given to changes in rain forests following destructive storms. A large storm may create an exceedingly large light gap in the forest, and the question arises whether post-storm succession is simply a very big light gap, or whether something qualitatively different happens because of the immense size of the damage done by the storm. This is an important question, and while the answer is still elusive, many studies are currently underway. A storm's damage to the forest is similar in many respects (though not all) to damage done by a logging operation. Understanding the natural processes of how the forest responds to a storm gap or a light gap might very well provide insight for designing more ecologically rational methods of forestry than those in current use.

A final point is worth making here. In the past it has been presumed by romantics that tropical rain forests are ancient places, superstable cathedrals of towering trunks whose age defies human imagination, hardly touched by the hand of *Homo sapiens* nor ravaged by the vicissitudes of nature. This vision is simply not true. Tropical forests are frequently subjected to tropical storms that periodically knock down almost all the trees, to landslides that remove large sections of vegetation, to naturally occurring fires, and most importantly in the past several thousand years (at least) to the hunting and agricultural pressure of *Homo sapiens*. Finding a truly "untouched" forest is hardly possible.

This is an important issue. Tropical rain forests can withstand a great amount of physical damage over the long term. They may appear fragile, but we now know they also have a high capacity to restore themselves—they may perhaps not be *resistant* to damaging events, but they are quite *resilient* if given the chance. They inevitably grow back after large storms, after landslides, and after peasant agriculture. The only damage that may be permanent results from modern degradations—urbanization and chemically intensive agriculture. A peasant who cuts down a patch of forest to plant corn for a couple of years and then abandons the site does little long-term damage (the forest grows back). But a banana company that physically alters the soil structure, chemically changes the content of the soil, and saturates the ecosystem with pesticides has a far greater effect. And a piece of land covered with cement will not recuperate as a tropical forest except over a very long period of time. We return to this theme in a later chapter.

Factor 6: Soils

Plants get most of their nutrients through their root systems. This means that the soil, where the roots lie, is one important determinant of how well a plant does. In understanding the basics of soils, it is appropriate to focus on two related but distinct questions. How does the plant get nutrients out of the soil? And how are nutrients stored in the soil?

How plants get nutrients out of the soil is well understood, largely because of the importance of this topic to the agricultural sciences. Many nutrients occur in the soil as small charged particles called ions. They are like lint particles with a static electric charge, except they are much, much smaller than lint particles. The most important nutrients that occur this way are potassium, magnesium, calcium, and one of two forms of nitrogen. All of these have a positive charge, and they are all attached to small clay particles, much like a small magnet attaches to a metal surface. As part of their normal life activities, plant roots give off positively charged ions (just as we must excrete urine, plants must excrete certain products of their metabolism). Positively charged ions create a high acidity in the soil (actually, the definition of acidity is effectively the ratio of positively to negatively charged ions). Thus, there is a continual gradient of acidity as you move away from the surface of the root. Because the plant roots give off positive ions, the zone immediately next to the root surface is more acid than the rest of the soil. When the clay particles with the positively charged nutrients attached come near to the plant roots, they encounter this more acid environment. Because the nutrients are now in a changed environment, they change their chemical form, and suddenly become available to be absorbed through the root. Metaphorically, it is sort of like the magnets attached to the metal surface are displaced by other magnets, the small positively charged ions that are excreted from the roots. And once the nutrients are free of the clay particles (once the magnets disengage from the metal surface), they are available to be absorbed into the root of the plant. It is important to note that it is the gradient of acidity that is important. As explained below, the positively charged ions are held fast in the soil at an equilibrium level associated with the acidity. It is the sudden change to a more acid state that makes them available to the plant root. The diagram in Figure 2.2 illustrates this process.

The way soils store nutrients is also well understood. Two parts of the soil structure are important: the clays and the organic matter. Clays are exceedingly large chemicals (tiny by absolute standards, but huge as chemical molecules go) whose surface is covered with negative electrostatic charges. Because of their negative charge they attract the positively charged ions in the soil. The greater the abundance of this clay component of the soil, the more positive ions will be attached. Here lies the key point. If the positive ions are not attached to anything, they tend to wash out of the system. Thus it is extremely important to have clay in the soil to keep nutrients from leaching out of the soil.

Humus, the decomposed organic material of dead plants and animals, functions in a similar way. Very small pieces of humus have negative electrostatic charges on their surface, and like the clay particles, they hold the positively charged ions on their surface. Thus, in terms of plant nutrition and soil fertility, humus acts exactly like the clays, attracting and holding onto positive nutrient ions in the soil, and then releasing them when it comes into contact with the plant roots. But organic matter has an additional use as well: as it decays it releases nutrient ions into the soil. It thus

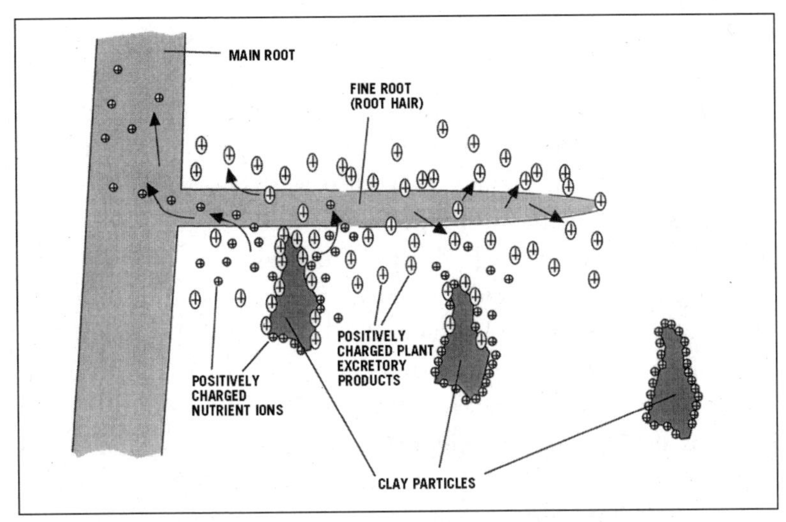

Figure 2.2. Diagram of how nutrients are absorbed by plants. A clay particle (or equivalently a piece of humus) carries with it small charged nutrient ions (small circles with pluses inside). These are the nutrients the plant needs for growth and development. The plant roots, especially the very fine tips of the roots, known as the root hairs, excrete other positively charged ions during the course of their normal life activities. These are the larger ovals with pluses inside them. As the clay particles come close to the plant root, some of the big positive ions replace the small positive ions on the surface of the clay, thus making those small ions available to be taken up by the plant through the root hair.

acts as a slow releaser of fertilizer.

Putting together the two concepts of nutrient storage and absorption, it is easy to see the fundamental mechanisms of soil fertility. To the extent that clay particles and humus in the soil have negative charges on their surface, they will tend to take up all the positive charges in the soil, including positively charged nutrient ions. But when one of these clay or humus particles moves close to the surface of a plant root, a more acid environment is encountered and some of their positive nutrient ions are replaced with positive ions that the plant has excreted. The positive nutrient ion is thus freed from the clay particle, and free to be absorbed through the root of the plant (see Figure 2.2). Note how important the acidity gradient is to the whole process. If, when a clay particle moves close to the plant root, it is not induced to give up some of the nutrient ions sticking to its surface, those ions will not be available to the plant. And if there is no gradient (for example if all of the soil is very acid so that the zone next to the surface of the root is not very different), none of those nutrient ions will be available.

That, then, is the general picture of soil fertility and plant nutrition. First, clay particles and humus hang on to the nutrients thus preventing them from washing out of the system. Then, when the clay and/or humus particles come within the vicinity of a root surface, they give up the nutrients they are carrying so the plant can absorb them.

Tropical rain forest soils have several characteristics that make them problematic.[12] First, they are highly acid. This means that the critical acidity gradient between the general soil and the root surface is less dramatic than in other soils. Thus any crops that farmers try to grow tend to have problems getting the proper nutrients from the soil. Second, rain forest soils usually have types of clay particles that bear few negative charges on their surfaces. They are thus not capable of storing very many nutrient ions. Thus, whatever nutrients may be present, even those from added chemical fertilizer, tend to rapidly leach out of the soil. Third, because of continually high temperature and abundant moisture, the process of decomposition occurs rapidly and the organic matter thus disappears rapidly when the natural forest is taken away. This means that both the source of nutrients and the storage capacity of the soil, consequences of soil organic matter, are very poor when farmers try to grow crops.[13]

The Diversity of Tropical Rain Forests

A fact often not fully appreciated in the popular imagination or literature is not only that tropical rain forests contain a great deal of biodiversity but that there is an enormous diversity amongst kinds of tropical rain forests as well. In a single region one may find distinct combinations of plants and animals on ridge tops as compared to valley bottoms, or on well-drained soils versus humid soils. Indeed it is the bane of the lumber operator in the American tropics that some patches of forest contain particularly useful species of trees, while other patches contain mainly species that have little market value.

Most important at this level of diversity is the distinction among major groups of forests. In Figure 2.1 we illustrated the location of the world's major tropical rain forests—the Americas, Africa, and Southeast

Asia. The forests in these areas do not operate by universal rules, and indeed sometimes differences among them are extremely important for the practical problems of forestry and agriculture.

For example, many of the forests of Southeast Asia are dominated by a single family of trees, the Dipterocarpaceae. In these forests this family may comprise up to 80 percent of the canopy trees and 40 percent of those in the understory, whereas in Africa there is only a handful of tree species that belong to this family, and in South America perhaps just one.[14] The Dipterocarpaceae family contains tree species that typically have straight trunks, grow to a very large size, and can usually be converted into valuable timber. Because of their generally large size and uniformly straight trunks, they are a timber company's dream. In a single hectare of Southeast Asian rain forest, one might encounter 200 species of trees, 180 of which will be in the family Dipterocarpaceae. By comparison, a similar plot in America or Africa might also contain 200 species, but they may be members of 50 different families, and the most common family might contain only ten species. Thus, almost any area of a dipterocarp forest can be utilized for cutting, as compared with the American or African tropics, where an area must be carefully scouted ahead of time to locate patches of valuable timber.

Furthermore, Southeast Asian forests have a very different physical aspect as compared with American rain forests. The understory vegetation in a Southeast Asian forest is composed mainly of small trees: the seedlings and saplings that will eventually grow up to become the large trees of the forest. In America, this vegetational stratum is dominated by plants that live only in the understory. This understory is where a wide variety of ornamental palms and other species come from. They never grow very large and live comfortably in a low-light environment such as is found in a rain forest understory, doctor's office, or suburban shopping mall.

Such differences suggest that the details of ecology are also quite distinct from place to place, and the more we learn of rain forests, the clearer it becomes that this is true. It is becoming increasingly difficult to write a chapter such as this, which attempts to make generalizations that apply to all tropical rain forests. Professional ecologists rarely refer any longer to "the tropical rain forest," but rather specify whether they are talking about an American, African, or Asian forest, and within those categories, they identify a myriad of subcategories.

The Biological Side Summarized

This chapter is a brief outline of some of the biological/ecological concepts necessary for understanding the rest of our book. It is useful to understand how a rain forest works, and certainly it will be helpful to comprehend both what happens, and why, when a rain forest is converted to agriculture. As will become clear in later chapters, knowledge of these issues will also be constructive as we try to come to grips with the difficulties of developing sustainable logging methods for rain forests. Finally, the political, economic, and social forces that are really behind the destruction of the world's rain forests are made clearer with the help of this background.

By now, hopefully, we have demonstrated that neither of our metaphors is perfectly accurate. Rain forests are not terribly fragile, nor are they especially stable. Rain forests are, however, highly complex and require a great deal of detailed plumbing and wiring and housekeeping—pollination systems, seed dispersal, defenses against herbivores, and proper soil conditions—in order to function properly. Rain forests may also even require periodic natural disasters such as storms and landslides, and they are attuned to the recurring formation of light gaps. And while the plants of the rain forest live in an Eden of heat and water, they suffer from a virtually empty pantry of the nutrients they need to survive.

3: FARMING ON RAIN FOREST SOILS

THE POPULAR PERSPECTIVE that the logger's chain saw causes rain forest destruction is incomplete. There are complex social and ecological forces involved in agriculture that are of far greater significance. Furthermore, sociopolitical factors are often of overwhelming importance. It is crucial to understand both the technical/ecological issues and the sociopolitical ones. In this chapter we present the relevant technical issues in a highly abbreviated form. In the two chapters that follow we cover the sociopolitical factors.

At the outset, we must acknowledge the temptation to assume that, in rain forest areas, the potential for agriculture is great. Since there is neither winter nor lack of water, two of the main limiting factors for agriculture in other areas of the world, it is easy to conclude that production might very well be cornucopian. The tremendously lush vegetation of a tropical rain forest only heightens this impression, and indeed this perception may ultimately be valid. Being able to produce for twelve months of the year without worrying about irrigation is definitely a positive aspect to farming in such regions. But, so far at least, the woes are almost insurmountable, as most farmers forced to cultivate in rain forest areas can attest.

The first problem is with the soils. As described in Chapter 2, rain forest soils are usually acidic, contain a type of clay that cannot exchange nutrients well, and are very low in organic matter.[1] Even if nutrients are added to the soil, they will be utilized relatively inefficiently because of the soil's acidity, and then they will be washed out of the system because of its low storage capacity.

This problem is actually exacerbated by the forest itself. Because tropical rain forest plants have faced these poor soil conditions for millions of years, they have evolved mechanisms for storing the system's nutrients in their vegetative matter (leaves, stems, roots, etc.). If they did not, much of the nutrient material would simply wash out of the system and no longer be available to them. This means that a vast majority of the nutrients in the ecosystem are stored in plant material rather than in the soil. This is exactly the reverse of the corn belt soils of the United States. Indeed the cycling of nutrients in a tropical rain forest is considered highly efficient, with surface roots absorbing nutrients as soon as they are released from decaying matter.

The consequence of this is that when a forest is cut down and burned, the nutrients in the vegetation are immediately made available to any crops that have been planted. It's like a bunch of fertilizer is applied all at once. The crops grow vigorously at first, but any nutrients unused during the first growing season will tend to leach out of the system. The "poverty" of the soil only becomes evident during the second growing season. This pattern is especially invidious when migrants from areas with relatively stable soils arrive in a rain forest area. The first year they may produce a bumper crop, which gives them a false sense of security. Then, if the second year is not a complete failure, almost certainly the third or fourth is, and the farmer is forced to move on to cut down another piece of forest.

A second problem is pests in the form of insects, diseases, and weeds. The magnitude of the pest problem is often not fully anticipated by farmers or planners, and it is only after problems arise that the surprised agronomists become concerned. This is unfortunate, since one of the few things we can predict with confidence is that when rain forest is converted to agriculture, many pests arrive. The herbivores that used to eat the plants of the rain forest are not eliminated when the forest is cut. They are representatives of the massive biodiversity of tropical rain forests, and the potential number of them is enormous. Herbivores can devastate farmers' fields, as they are able to destroy an entire crop in days.

That same uniform moist and warm environment that suggests high potential for agriculture in rain forest areas is also quite hospitable for the organisms that cause crop diseases. Consequently the potential for losing crops to disease is far greater than in more temperate climates. And just as the hot and wet environment is agreeable for crops, it is also agreeable for competitive plants. Since no two plants can occupy the same space, frequently the crop falls victim to the more aggressive vines and grasses that colonize open areas in tropical rain forest zones. Weeds are thus an especially difficult problem.

Rain Forest Landscapes: The Invisible Mosaic

Like other ecosystems, tropical rain forests exist in diverse settings. These settings frequently set the stage for the details of agricultural incursion into an area. In particular, the nature of the soils is governed by the same forces that govern the formation of the landscape. In any rain forest area we can expect to find a variety of basic soil types. There are five basic soil types of relevance to agriculture in rain forest areas: acid soils, alluvial soils, volcanic soils, hillside soils, and swamp soils.[2]

Acid soils are the classic rain forest soil. When the main bulk of the soil is acid, the acidity gradient between the general soil and the area immediately adjacent to the plant root virtually disappears. Since this gradient is essential for plant nutrition,[3] acid soils are very poor when it comes to feeding plants. Because of this acidity, even adding nutrients to such soils does not make much difference. Furthermore, acid soil usually contains a very small amount of organic matter and the sort of clay minerals with low negative charges. Thus, the nutrient ions which do get into the soil are rapidly leached out. In sum, acid soils contain few nutrients to start with, and those contained are frequently unavailable to the crops anyway.

The origin of these acid soils is not mysterious. When a soil first comes into being, the clay particles[4] it contains are large and incorporate a great deal of aluminum and iron within their chemical structure. The large surface of each clay molecule provides many sites for negative electrical charges. These negative electrical charges attract positive ions dissolved in water in the soil, thus making the soil less acid[5] and keeping the positive ions around where they can later be used as nutrients by the plants. But weather effects, especially high temperature and moisture, cause the clay molecules to decay. As they decay, they become smaller. As they become smaller, there are fewer sites for negative charges, and some of the aluminum and iron that was contained within the clay molecule is released to the soil in the form of ions.[6] Since these are positive ions, they contribute to increasing the acidity of the soil. Because the smaller clay molecules can no longer neutralize as many positive ions, this also contributes to the soil's becoming more acid. This process, known as weathering, is what makes acid soils.

Alluvial soils are deposited on the floodplains of rivers. Because of the abundance of living creatures in rivers, alluvial soils tend to carry a lot of organic matter, which gets deposited on land during floods. The mud left after a river floods may look ugly and do immense damage to a living room, but it contains large quantities of biological materials. Because of this organic matter, alluvial soils can store nutrients quite well.[7] The Mississippi and Rhine river valleys are excellent examples of areas with rich alluvial soils, rightfully famous for their agricultural productivity. Wherever there are rain forests, one can almost always find alluvial soils along the borders of the rivers that flow through them. One can frequently draw an approximate map of where the alluvial soils are, simply by marking all the agricultural fields, as farmers can be quite good at finding the best soils in an area.

The third major soil type is derived from volcanic ash. Volcanic soils do not always occur in rain forest areas. Indeed they are an exception rather than the rule, but when they do occur they are important agriculturally. They are usually extremely productive, sometimes rivaling the rich corn-belt soils of North America. As they age, they too become acid through the weathering process. But such weathering requires hundreds or thousands of years, and while they are young, volcanic soils do not have the undesirable qualities of acid soils. Young volcanic soils are particularly good for agriculture because the clays they contain are well known for their ability to capture and retain nutrients. In some of the rain forest areas of Java and Hawaii, for example, volcanic soils are very fertile. The beautiful Javanese rice terraces would likely still be rain forests had not the rich volcanic soils led farmers hundreds of years ago to cut the forest and replace it with rice fields. For obvious reasons, rain forests growing on these rich volcanic soils throughout the tropics were the first to be cleared for agriculture.[8]

Hillside soils have one important characteristic: they erode very rapidly. The natural vegetation that covers them is effectively the only protection they have against severe erosion. When converted to agriculture, hillside soils are rapidly eroded; the topsoil washes away and the land soon becomes unproductive. Unfortunately, because of economic and sociopolitical pressures, many peasant farmers are forced to farm on these soils, with inevitably poor results.

Finally, many tropical areas encompass natural wetlands, whose soils (in the U.S. they are called swamp or muck soils) are extremely rich in organic matter. However, except for rice and a few other crops that can grow under inundated conditions, it is not possible to do agriculture in a swamp. But with large-scale projects, swamps can be drained, and the resulting soils are extremely rich. This, however, is not an activity normally undertaken by small farmers. Rather it is more likely to be associated with larger, capitalized operations like those of the big fruit companies.

In light of the above, it is oversimplification at best to state that all tropical soils are poor. There is a range of fertility associated with tropical soil types that depends upon local conditions. From the rich volcanic soils, to the moderately rich alluvial soils, to the rich but hard-to-manage swamp soils, to the extremely poor acid soils, to the hillside soils that erode so easily, there is inevitably a mosaic of soil types, and consequently a mosaic of agricultural potentials, which exists in rain forest areas. Note the distribution of soil types in an imaginary landscape that could arise as we depict them in Figure 3.1.

Without knowing the history of such an area (that there is an old river channel, that part of it is covered with an old volcanic ash deposit, etc.), one would be hard pressed to know exactly where the good and bad soils were located. There is an unseen mosaic of soils underneath any rain forest, some favorable for agricultural production, others less so. In Figure 3.2 we present a production plan based on this soil mosaic. However, without a detailed study, something the peasant agriculturalist is not likely to

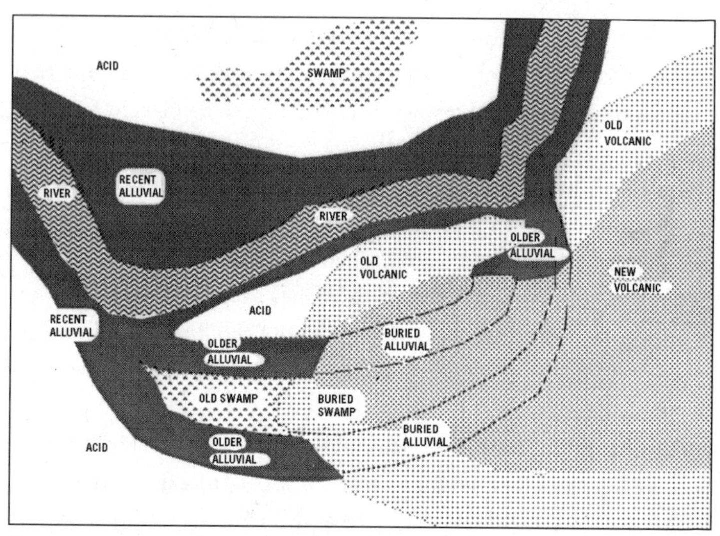

Figure 3.1. Hypothetical illustration of typical distribution of soils under a tropical rain forest. The river running through the forest deposits the recent alluvium every time it floods. The river used to have a different channel and used to deposit alluvium in that channel before natural erosion changed its course. This is what made the old alluvial deposit. Part of that older alluvial deposit was buried by a recent volcanic eruption, and thus is called buried alluvial. An earlier volcanic eruption deposited ash in a larger area, and before the river changed its course. Thus the older alluvial soil is actually on top of the soil that was produced from the earlier volcanic eruption. The soils that were never covered with either volcanic ash or alluvial deposits are the oldest soils and are very acid. A small swamp used to exist just north of the river, creating a humus-rich swamp soil. From a farmer's point of view the soils range from the really excellent young volcanic soil to the recent alluvial, to the swamp, to the old volcanic, to the old alluvial, to the acid. At one extreme (new volcanic) the soils are quite good, but at the other extreme (acid) the soils are among the worst in the world. And the trick is that you can't necessarily tell where all of these soils are since everything other than the river is covered with rain forest.

be able to afford, one cannot tell where those good soils lie, and our plan, however rational ecologically, would be impossible to formulate under current political conditions. Thus, most actual agricultural planning in rain forest areas is done in relative ignorance of the underlying soil mosaic. This problem is faced by both the small farmer and the planner who seeks to promote efficient agriculture in such areas.

Yet even though this unseen mosaic is a reality in all rain forests, it remains true that a large proportion of the soils under current rain forest

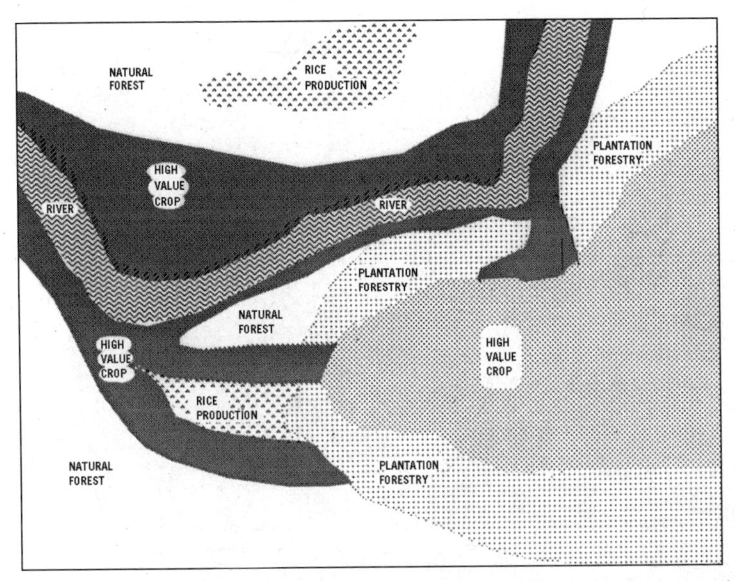

Figure 3.2. A production plan based on the underlying soil mosaic. Each cropping system is based on the land-use capability. Note how such a system would not conform well to the notion of straight-line fences and rows.

cover are of the highly acid type. Indeed, in many areas of the world, the volcanic and alluvial soils have already been converted to agriculture and only the acid soils and soils on very steep slopes remain covered with forests.

Slash and Burn Agriculture

The most common form of traditional agriculture, and one still practiced in many places in the world today, involves the use of fire and a reliance on the ecological process of succession (see Figure 3.3). In principle, slash and burn agriculture (frequently referred to as shifting cultivation) represents the easiest solution to two ecological problems inherent in agricultural production: the problem of plant competition (weeds), and the problem of nutrient cycling. If crop seeds are scattered in a patch of cleared forest, the plants already there will have a clear advantage over the newly seeded ones. By burning the cleared patch before scattering the seeds, the farmer forces the undesirable species to at least begin at the same level as the crops.

Figure 3.3. Slash and burn agriculture. Top photo is of a recently slashed and burned area in Tabasco, Mexico. Bottom photo is the same area after crops have been growing.

The second ecological problem is nutrient cycling. Most of the nutrients in a rain forest are recycled very rapidly. As dead leaves decompose, they release their nutrients into the soil, where they are then picked up by the roots of living plants and reused. We have reason to believe that most systems unperturbed by *Homo sapiens* are in relative balance regarding nutrient cycling—the amount of nitrogen, for example, released into the soil by decomposing leaves equals the amount of nitrogen picked up by the trees as they put on their new leaves.

But consider what happens when agriculture is imposed on the system. Harvesting means removing some of the biological material from the system. It is not then available for recycling. Burning exacerbates this problem tremendously in that nutrients normally stored in plant material—thus held securely in the system—are released en masse by the burning process. Crops utilize some of these nutrients, but the rest may be taken out of the system through water runoff or leaching. After a while, since the nutrient cycling process is interrupted and there is a net export of nutrients out of the system, the land must be abandoned.

Subsequent to abandonment, the process of ecological succession takes over and various plant and animal species occupy the abandoned agricultural land, slowly bringing it back into a relatively stable, nutrient cycling system. After a fallow period during which the patch has returned to forest, the farmer may return, slash and burn the new vegetation, and start the cycle over again. Depending upon the nature of the soil, and of the successional process that takes over after the land is left idle, this cycle may be as short as ten years, or as long as fifty. In a sense, ecological succession "solves" the problem of the disruption of the nutrient cycling system caused by agriculture.

The slash and burn system can be seen as cutting and applying fire to natural vegetation to release the desired plants (the crops) from biological competition, and to provide them with large amounts of nutrients. However, slash and burn creates a problem of net nutrient loss from the system, which requires abandonment of a particular site after a short period of time (usually two to ten growing seasons) to allow the process of ecological succession to bring the system close to its original pattern of nutrient cycling.

Stabilizing Agriculture on Rain Forest Soils

Before beginning this section, we wish to remind the reader that the reasons most small farmers move their farms into rain forests have much more to do with politics than with nature, a point to which we return in the next chapter. But even if the sociopolitical landscape were utopian, the biophysical conditions of rain forests make them difficult to farm in perpetuity. At least that is the case with current forms of peasant agriculture. However it is also true that some indigenous people have farmed within rain forests for generations without major disruption of the overall forest landscape.[9] Sometimes these indigenous populations are very small and practice slash and burn agriculture at the margins of old rain forest, allowing their fallow only ten or twenty years of succession before returning to it. More rare are those groups who actually farm within an intact forest. We know of few such examples.[10] Most instances of indigenous people "farming" the rain forest are really examples of shifting agriculture in which the older fallow looks much like an old growth forest. While the lessons we may gain from such systems are important and potentially useful to nontraditional people, we emphasize that such populations represent a very small fraction of the agriculture currently practiced on the margins of rain forests.

Given the above, it is important to ask what might be done, technically, to stop peasant agriculture from incursions into rain forest areas. That is, even in the bright hopeful future when small farmers have full title to their land, the full security of credit, and the availability of fair markets,[11] the technology currently available for producing basic grains on rain forest soils is not sustainable. The soils become depleted of their nutrients rapidly, weeds build up to unacceptable levels, and frequently pests and diseases become uncontrollable, forcing the farmer to move onto a different piece of land. Ultimately we must deal with this problem.

The first and most obvious solution is simply to restrict agriculture to the patches of relatively good soils that exist in rain forest areas. To a great extent these were precisely the areas cleared for agriculture in the first place. For example, in some areas of Central America one can almost draw a map of where the patches of good soils are located, simply by mapping the banana plantations. These plantations have been placed, for the most part, on alluvial soils. A quick glance at Figure 3.1 suggests that if we had a rational society we would designate the new volcanic and the recent alluvial soils for agricultural purposes, and probably devote the rest of the soils to forestry or simply let them remain as natural forest, as suggested in our action plan in Figure 3.2. But in a time when rational societies are apparently regarded as utopian, we must look for more practical solutions. We are thus left with the problem of farming on acidic rain forest soils.

Based on proven technology, there are three general principles that should be followed in planning agriculture in tropical rain forest areas. First, do not attempt basic grain production on highly acidic soils.[12] Second, grow wood (trees) on the poorest soils. Third, incorporate perennial crops (crops that do not die after a single year of production, such as cooking bananas or fruit trees) in whatever agricultural activities are proposed. These points require little elaboration. Except for the traditional method of slash and burn, which requires a large amount of land per person, the prospects for productive agriculture on rain forest acid soils are not great. Future technologies may change this assessment, and someday we may be able to take advantage of the twelve-month growing season and abundant water supply of tropical rain forest zones. However, with today's technologies the prospects are quite dim. The ecosystem that best develops on acid soils is rain forest, with trees as its main structural component. Ultimately we expect it will turn out that growing trees on acid soils is the best use to which such soils can be put.

Agricultural activities begin to look more attractive in the patchwork of old swamp soil, old alluvial soil, old volcanic soil, or any other soils with patches of less acid (and therefore potentially better) soil. It is here that some of the successes of intensive agricultural production occur in rain forest areas. Examining these successes, one is struck by a single common feature: they are almost always based on perennial crops, or have perennial crops as part of the overall system.

For example, Daniel, a farmer we know, has been farming the same patch of land in Nicaragua for twenty-five years. His corn production is based on a two-year cycle, with a fallow year between each production year. The fallow is largely dominated by two particular species of plants, a small bananalike herb known as wild banana and a morning glory vine. Daniel argues that these plants are an important part of the overall system since they keep the grasses out. According to Daniel, the grasses "burn the corn" (by which he means they are bad weeds that reduce corn production). If a patch of fallow starts showing signs of grass, he immediately cuts the grass out by hand. When he is ready to plant the corn, he simply cuts down the morning glory vines and the wild banana, and plants the corn amidst the mass of dead leaves and stems. As the corn grows, so do the vines and wild bananas, presumably choking out the grasses which would "heat up the corn" too much.[13] Amidst Daniel's farm is a virtual forest of fruit trees. Coconut, mango, cashew, and a host of other fruit trees dominate the agricultural system. The corn production is accomplished in the gaps within a forest of fruit and timber trees that dominate the overall agroforestry system. When Daniel talks about his farm, he mainly talks about the fruit trees.

Similarly the celebrated Javanese home gardens are famous for the high diversity of fruit and other trees that inevitably form the backbone of this production system. The gardens are small areas (usually less than one hectare) of mixed trees and crops adjacent to the house. They are characterized by a multistoried vegetation structure and a high diversity of plant species, resembling the configuration of a tropical forest. More than 600 species of plants are known to be grown in these home gardens.[14] They normally include staple foods such as taro or cassava, a variety of fruit trees and vegetables, spices, medicinal plants, herbaceous plants, trees used for building materials and firewood, and, finally, cash crops such as coffee, cacao, pepper, oil palm, and even some timber species such as teak. Many home gardens also have domesticated animals such as chickens and pigs, as well as fish grown in ponds.[15] Given the variety of food and fiber produced in these home gardens, it is easy to appreciate their importance for the nutrition of Javanese families. Although home gardens are most conspicuous in Java, occupying 15 to 75 percent of the cultivated land,[16] they can also be regularly found in other tropical areas of Asia, Africa, and Latin America. They are always located on very small parcels of land, intensively managed, with a wide diversity of plants and animals. But the key feature is the predominance of trees in the system.

Even in one of the most notorious agroecosystems in rain forest areas, the predominant crop is a perennial. The banana plantations of Central and South America are based on a perennial crop. While all the banana stems are cut once per year, they are not killed, and each banana stem resprouts into a new tree. Thus the basic idea of maintaining a structured root system in the soil is an essential feature of even this high-tech system. Other cash crops now infamous in tropical rain forest areas include cacao (chocolate), rubber, oil palm, and a host of tropical fruits, all perennial crops.

The success of agroecosystems based on perennial crops is not yet fully understood, but one simple explanation makes intuitive sense. Since the main problem is nutrients being exported from the system during times when no crops are covering the ground, the placement of permanent plants in the soil provides a network of roots to "catch" the nutrients before they run out of the system. When annual crops are the main feature of the system, the soil is bare for part of the year, allowing the abundant rainfall to percolate through. As it percolates, it washes away the nutrients. With living roots permanently in the soil, much of the water is absorbed, along with the nutrients dissolved in it, into the plant tissue and thus does not simply wash through the soil. The net result is that perennial crops tend to reduce the loss of nutrients from the system.

Some Models for the Future

For the immediate future we see little hope for major advances on the technical front. What is easily knowable seems already known, if not obvious: use of perennial crops and trees, and restriction of agricultural activities to the patches of relatively good soils. Since many farmers who are currently forced to farm in rain forest areas are not utilizing these obvious methods, a great potential exists for improvement, with little or no technological development. For the most part it is a question of political will, a topic to which we return in later chapters.

On the other hand, some intriguing theoretical models are available. Gómez-Pompa, for example, based an ideal model of agriculture in lowland tropical Mexico on the presumed production technology of the ancient Maya.[17] The Chinampa system, widely used by the Aztec cultures in highland Mexico, is a complicated integrated system of agriculture and aquaculture, based on raised fields and the physical manipulation of large amounts of organic matter. To the extent we understand how the Maya used what appears to be a similar system in the lowland tropics, it represents a system of tight nutrient cycling in rich swampy regions.

In Figure 3.4 we summarize the operating principles of the Chinampa system—at least those that remain functioning in the central valley of Mexico today. Terrestrial platforms are constructed from swamp soils, creating a network of canals. Biological activity in the canals (photosynthesis, herbivory, and decomposition) utilizes the nutrients in the water and incorporates them in biological material in the form of aquatic plants, fish and other aquatic animals, and the rich organic muck at the bottom of the canal. The fish and other aquatic animals can be harvested directly for human consumption. The aquatic plants can be harvested and used as a mulch on the terrestrial platforms, or composted and incorporated into the soil as organic matter. The organic muck at the bottom of the canal is used to form seedbeds. Agricultural production on the terrestrial platforms thus utilizes the materials from the aquatic ecosystem to improve the quality of the soils.

But the key to the system is the nutrient runoff problem. The Chinampa system solves this problem elegantly, not by trying to stop nutrient runoff, but by expanding the system to include a part that literally catches the nutrients as they run off. The aquatic vegetation utilizes the nutrients that escape the terrestrial agricultural system and reincorporates them into the biological material. The farmer then physically returns those nutrients to the agricultural component of the agroecosystem. Note that the "agriculture," the part of the system that actually produces the crop, is only a part of the "agroecosystem." The agroecosystem includes both the agricultural platforms and the canals.

Another example, not currently in use in a lowland rain forest region, is the soil management practices of the Mayans in the Almolonga Valley of Guatemala. Here farmers utilize the leaf litter of the forests in the surrounding hills as an organic mulch and incorporate it into the soils as they prepare the land for planting. In those hills one can find various combinations of oak, pine, and tropical forest vegetation. Depending on the crop and the perceived condition of the soil, various formulas for combinations of litter are utilized. Production appears to be high and sustainable.[18]

One of the most important features of the Almolonga system is the conception the farmers have of their production unit. The agroecosystem

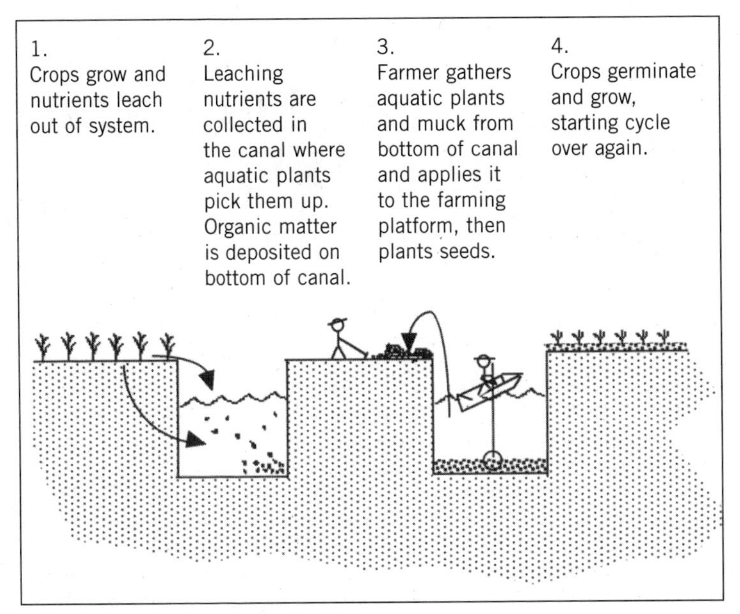

Figure 3.4. Functioning of the Chinampa system, as described in the text.

is not simply the fields in which they produce their crops. The Almolonga farmers conceive of their agroecosystem as including the forests on the surrounding hillsides, as much as the valley bottom land where their fields are located. While the trees may not be located exactly in the fields, their presence is recognized as extremely important to the function of the overall system.

Unfortunately these examples are quite theoretical. Although both are functional agroecosystems, they are not located in lowland wet tropical areas. While the archaeological evidence is relatively clear that the Chinampa system was utilized in one form or another by the Classic Maya, it is no longer in use, and we really do not know the details of how the Maya actually managed it. Despite valiant efforts, thus far attempts at reconstructing Chinampa systems in tropical areas have not met with much success,[19] and we are left with the unfortunate conclusion that both the Chinampa and Almolonga systems are simply intriguing models that someday may form part of the solution to the problems of agriculture in rain forest areas.

4: THE POLITICAL ECONOMY OF AGRICULTURE IN RAIN FOREST AREAS

AS DESCRIBED IN CHAPTER 1, during their normal operations loggers open up roads, cut many of the larger trees, and do extensive secondary damage to rain forests. Though there is evidence that tropical rain forests are able to withstand this sort of intervention, that capability is significantly reduced when the forest is converted to agriculture.[1] The most damaging kind of agriculture for forest regeneration is the modern capital-intensive system. To understand this system fully, it is essential to learn how it came to be and how it functions. In this chapter, we trace the development of agriculture from its early origins to its current modern export base. In the chapter that follows we discuss how agriculture operates today, in both the developed and the underdeveloped world.

We begin with three separate but related topics: (1) the origin and intensification of agriculture; (2) the origin and spread of enclave production, a form of agricultural social organization important in the world's tropical regions; and (3) modernization, especially as it relates to machines, chemicals, and food processing.

The Origin and Intensification of Agriculture

In the previous chapter we touched upon the slash and burn system, emphasizing its ecological properties. Undoubtedly the first well-developed form of agriculture, slash and burn probably appeared for the first time in the Mediterranean. While this system is usually practiced in savanna or scrublike vegetation, it was also practiced in tropical rain forests occupied by human beings. Thus, slash and burn agriculture represents the first significant human activity that can be thought of as "deforestation."

As with all human practices, slash and burn agriculture is continually evolving. One pattern of that evolution is a tendency to shorten or lengthen the fallow cycle (i.e., the period when the plot is left idle and secondary succession occurs). Thus, a twenty-year cycle might evolve into a sixteen-year cycle and then a ten-year one, or it could evolve in the other direction into a twenty-five, and then a thirty-year cycle. When fallow time is shortened, it is known as "intensification" of agriculture. "Extensive" systems involve large amounts of land in fallow and gradually give way to "intensive" ones in which all the land is simultaneously under cultivation. However, as explained below, this is only one possible pattern.

Earlier agricultural analysts emphasized a sort of Malthusian approach to this topic, suggesting that it was necessary to put increasing amounts of land into cultivation in order to feed an ever-growing population, thus leading to an inevitable shortening of the fallow cycle. There was thought to be a simple relationship between agriculture and population growth, with population growth the underlying driving force, and increased agricultural production allowing population to reach its more-or-less natural high level.

This view was discarded some time ago. The labor required to make land agriculturally productive is large, and frequently it is agricultural production itself that promotes a faster rate of population growth, precisely the opposite of what had previously been assumed. If a farmer sees that a certain number of acres must be planted to feed her family, she understands that a certain number of workdays are required to make that piece of land productive. If the number of workdays is considerable, it is natural to hope for sons and daughters to help with those hours. Population growth is thus promoted. That these sons and daughters will some day require yet more land to feed them is not likely to come to mind during the long hours of backbreaking work in the field.[2]

However, it is almost certain that the opposite occurs also. The labor necessary to feed a family may be small due, for example, to a small family, or highly productive soil. The farmer will not need to work many hours. The tendency then will be in the direction of overproduction. There will be a surplus of food, indicating that so much land in production is unnecessary, and the next year the amount of land cultivated will be reduced. This surplus may arise for various reasons. For example, a climate change may result in higher production, or some of the population may emigrate, lowering overall food requirements.

Thus we have a pair of countervailing tendencies. On one hand is a tendency to have more children, or to attract migrants to help work the amount of land necessary to support the local population. This means the population will increase, requiring more land to be put into production and therefore more labor to work the extra land—a continual cycle. On the other hand is a tendency to take land out of production when a surplus of food is produced. The need for more field labor is reduced, and thus does not provide a particular incentive to have more children or attract migrants. It could even result in a decreasing local population if some force from outside the region were acting as a magnet for emigration.

These countervailing tendencies can produce an increased population, a decreased population, or a stable one. The same tendencies may also result in increasing, decreasing, or stabilizing the amount of land in production. The point is that there is no necessary tendency for a population to grow, require an ever-increasing amount of food, and thus necessitate more land under cultivation. That may be the case under some conditions, but others may provoke exactly the opposite result, or even a stable population balanced between the desire for a larger population to help work the land, and a tendency towards a smaller population due to a deficit of production.

Once the full intensification of agriculture occurs (i.e., once there is no longer any fallow period), a variety of new tendencies begin to appear. Under a slash and burn system the easiest way to respond to social and biological events is either to create more fallow land or to take more land out of fallow, giving rise to the sorts of changes described above. But what sorts of responses would be expected if all available land had already been taken out of fallow? The same pressures would likely be felt, but the solution of modifying the fallow period would no longer be available. At this juncture agriculture begins the modernization process.

Recall that burning the field was the major form of land preparation, the main purpose of which was to remove competitive vegetation—weeds. But much of the weedy material would have underground stems that would send up shoots and rapidly become a problem after burning. Thus, while burning certainly removes weeds, it does not do so completely. Probably in response to the persistent problem of weeds, the idea gradually developed of digging up the underground parts of the weeds, ushering in the era of the plow.

The plow would not have been of much use without draft animals to pull it—and it is likely for this reason that the plow never evolved in the Americas.[3] The plow as a major innovation generated other innovations, especially those involved with metallurgy. Thus the Bronze and Iron ages of the Old World never materialized in the Americas, possibly due to the lack of this all-important technological innovation. But apart from the role of agricultural modernization in shaping other aspects of world history, the plow initially led to the evolution of various agricultural implements that made the farming enterprise ever more effective and finally led to the complete modernization of agriculture.

Before we describe the modern agricultural system in detail, a very special form of agriculture, "enclave" production, must be considered. While marginal in the current modern world, "enclave" production is sometimes closely connected, both directly and indirectly, with the deforestation of tropical rain forests.

Agriculture and Enclave Production

There is nothing more British than teatime, yet that custom is an outgrowth not of ancient kings, dukes, and earls, but rather of British colonialism. Tea cannot be produced on the British Isles, and it was not part of the culture until the colonies in South Asia were formed. Not trusting local people to take responsibility for anything as important as profits, the English East India Company evolved a form of production that proved to be extremely efficient, albeit at the expense of the people of India and the other colonies.

The key idea was to integrate all aspects of production, even very indirect ones. This enabled the British to maintain control over each component of production, from planting the tea bushes, to harvesting the leaves, to drying and packaging the tea, to shipping. Workers were both permanent and seasonal, but, as much as possible, were housed in company-owned barracks and bought their provisions at company-owned stores. British employees hardly knew they were in India since all their needs were met on the company compound. It was as if there was a little Britain in the Indian highlands, which is why the system came to be known as enclave production. Furthermore, this sort of production can lead to major political influence, as witnessed by the central role played by the English East India Company in the history of Britain in South Asia.

The major remnant of this kind of production is banana production in Central and South America, which began quite differently than did the tea enclaves of India, but whose consequences have been just as devastating for both the social and the physical environment of the region. Much as the British enclave production in Asia led to the growth and concentration of British political power in the seventeenth and eighteenth centuries, so the international political scene in the American neocolonies of the United States was strongly influenced by the companies that controlled the enclave production of bananas. How was it that such power and influence arose?

Late in the nineteenth century, Lorenzo Dow Baker, a U.S. ship captain, picked up a small load of a curious yellow fruit on a stop in Jamaica. Upon arrival in Boston he sold the load so rapidly that he immediately saw an immense future potential for this new agricultural product, and teamed up with Andrew Preston, an entrepreneurial friend, to form the Boston Fruit Company. Baker purchased the fruit from producers in the Caribbean to bring to Boston, and Preston was in charge of marketing in the U.S. The company flourished.[4]

Along about the same time, Minor C. Keith was looking to fulfill his destiny as a railroad man. However, demand for new rail transport in the United States had declined, and Keith faced the possible failure of the fulfillment of his mission in life. He moved to Costa Rica, where he obtained a government concession to build a railroad from the central city of San Jose to the port town of Limón. Having built the railroad, he soon found that Costa Rica's economy was not exactly dynamic, and that there was really very little for his new railroad to do. But just about at that time he heard of the remarkable success of the Boston Fruit Company, and realized that Costa Rica had a perfect climate for banana production. Keith saw that bananas could give his railroad its raison d'etre, and got into the banana business.[5]

The banana trade grew rapidly, with Guatemala and Honduras joining Costa Rica as major production sites, and as many as ten companies producing or buying bananas and eight companies shipping them. Then, in 1898, the Boston Fruit Company teamed up with Keith's operation and purchased the major shipping and production companies involved in the banana business, forming the United Fruit Company. Guatemala, Honduras, and Costa Rica have not been the same since.

The United Fruit Company grew by leaps and bounds over the next half century. By the early 1950s it was the largest landowner in Guatemala, and one of the largest in both Honduras and Costa Rica. Unlike other foreign ventures that established businesses that integrated with the rest of the national economy, the United Fruit Company maintained the classic enclave production philosophy, as invented by the British. Every aspect of production remained under company control. Along with its control was an ability to rapidly expand if conditions were auspicious. United Fruit accumulated a great deal of land. Not only did it have a huge amount of land in production, it also held an equally large amount of land in reserve, possibly for speculative purposes, or perhaps for future banana production.

Export agriculturalists often purchase or even steal land from peasant farmers situated on good soil, forcing the farmers onto poorer soils on the hillsides or in the rain forests. The United Fruit Company in Guatemala is an excellent example.[6] By 1950, United Fruit owned 565,000 acres of land, making it the largest landholder in Guatemala. At the same time 75 percent of the peasant families were either without land at all, or had small plots of land that were marginal at best. Such an arrangement was clearly tinder for a social explosion.

The social explosion came, not as a violent outburst, but rather as a sophisticated and civilized political campaign. In 1950 one presidential candidate, Jacobo Arbenz, ran on a ticket of agrarian reform. He promised to purchase land held by large landowners and redistribute it to small peasant farmers. Arbenz won the election on that platform, and his agrarian reform program was rapidly put into action. On farms of more than 223 acres, land not being used for production was targeted for expropriation. The expropriations were made up of lands taken from 1,059 farms with an average size of 4,300 acres. Approximately 100,000 peasants received title to the land thus expropriated. It is not difficult to see why Arbenz was popular.

Probably Arbenz's undoing was the expropriation of United Fruit Company land. While the United Fruit Company was the largest landholder in Guatemala, only 9 percent of its land was actually being used. Consequently, about 240,000 acres on the Pacific coast and 173,000 acres on the Atlantic coast were expropriated, for which the company was compensated with $6,000,000, based on its own stated value of the land.

Arbenz had challenged the basic arrangement of the Guatemalan economy.[7] To the politicians and their clients in the United States of America, such a challenge seemed a harbinger of things to come and caused a great deal of consternation. If it was true, as the president of Castle and Cook (then the parent company of United Fruit Co.) put it, that "Without overseas investment your entire economic infrastructure collapses in on you,"[8] then Guatemala's taking charge of its own resources was certainly a threat to free-wheeling investment opportunities available in the Third World. And if Guatemala, as one of several dozen countries that performed the same function, was not particularly important in and of itself, what might happen in Costa Rica or Nicaragua or Honduras if Guatemala were successful? Indeed, shortly after Arbenz's victory and initial confrontations with the United Fruit Company, both Costa Rica and Honduras began making demands of the United Fruit Company, presumably buoyed by Arbenz's success.

Additionally, the United Fruit Company was a powerful entity not only in Guatemala, but in the United States. As the owner of over 300 million acres of land, 2,000 miles of railroads, and 100 steamships in Central America, the United Fruit Company was an influential player in international politics. Furthermore, U.S. secretary of state John Foster Dulles was a senior partner in the Sullivan and Cromwell law firm, the principal legal agent for the United Fruit Company. His brother Allen was the head of the Central Intelligence Agency (CIA).

Finally, and most importantly, at about this time, the cold war was becoming a defining feature of U.S. politics. The Eisenhower administration (especially Vice President Nixon) was to set the stage for the cold warriors.[9] The international Communist movement advocated quite a theoretical package for the oppressed, and it was not difficult to see Communists behind every threat against a capitalist venture wherever it occurred. Thus, despite the evident non-Communist nature of Arbenz himself, the fact that his legislature contained several deputies who were members of the Communist party, and the simple fact that he promoted agrarian reform, either led the cold warriors to believe there was a Communist conspiracy afoot, or at least enabled them to justify their actions based on the claim of a Communist conspiracy.

In 1954 the Eisenhower administration went into action. A lesser colonel in the Guatemalan army, Castillo Armas, was chosen by the CIA to "lead" the revolt against the Guatemalan government. The plan had been hatched almost a year earlier at the upper levels of the U.S. government. The CIA knew that neither the Guatemalan people nor the Guatemalan military would rally to the cause of overthrowing the popular Arbenz, so they had to rely on a strategy of causing Arbenz to surrender without a true military threat. This meant that they had to create the impression of a real threat to convince Arbenz that he had no way of avoiding a bloodbath other than resigning. Through some duplicitous diplomacy and a great deal of propaganda, the Arbenz government was led to believe that Castillo Armas had a large force (he never actually had more than 300 troops) poised to engage the Guatemalan Army, and that U.S. armed forces were ready to enter the fray in support of the overthrow. Most spectacular was the bombing of Guatemala City by CIA planes, generating panic in the city, as planned. Arbenz, fearing a bloodbath, capitulated, following almost exactly the script the CIA had plotted.[10] And, most importantly for rain forests, the agrarian reform program was abandoned.

Today, despite retaining its basic enclave form, the modern mode of banana production is technically far more sophisticated than it was in the days of Arbenz. To establish a modern banana plantation, it is often neces-

Figure 4.1. Banana production in Central America. Photo on top shows the metal monorail that runs throughout the plantations. Photo on bottom shows banana workers harvesting a bunch of bananas. Photos are from the Standard Fruit Company plantations in Rio Frio, Costa Rica.

Figure 4.2. Aerial view of the United Fruit Company banana plantation in Costa Rica.

Figure 4.3. Banana workers prepare metal monorail to receive harvested bananas to bring them to the packing plant.

sary to construct a complex system of hydrological control wherein the soil is leveled and crisscrossed with drainage channels. This significantly alters the physical nature of the soil. Contemporary banana production includes burying plastic tubing in the ground to eliminate the natural variability in subsurface water depth. Metal monorails hang from braces placed into cement footings to haul the bunches of bananas (Figure 4.1). To avert fungal diseases, heavy use of fungicides is required, and because of the large scale of the operation (Figure 4.2), chemical methods of pest control are the preferred option. The banana plants create an almost complete shade cover and thus replace all residual vegetation. Pesticide application is sometimes intense, other times almost absent, depending on conditions, but over the long run one can expect an enormous cumulative input of pesticides, the long-term consequences of which are unknown but unlikely to be salutary.

A major social transformation is also required to set up a modern banana plantation. The factorylike conditions of production require a semi-proletarianized work force, sometimes drawn from the local peasant population, but usually brought in from surrounding areas (Figure 4.3). Banana production tends to promote a local "overpopulation crisis" by encouraging a great deal of migration into the area in which it operates. As the international market for bananas ebbs and flows, workers are alternatively hired and fired. There is little alternative economic opportunity in banana zones, and fired workers must either look for a piece of land to farm, or migrate to the cities to join the swelling ranks of shantytown dwellers.

The Modernization of Agriculture in the Developed World

The popular understanding of agriculture in a country like the United States is sometimes hindered by an overly romantic notion of the farm. As part of the original expansion of European America and something of a centerpiece of American-style democracy, the image of the rugged farmer carving a homestead out of the wilderness is an important piece of our national lore. While the small family farm may have been important in both the Europeanization of America and the establishment of the form of democracy practiced earlier in our history, the farm and farmer of today bear little resemblance to this romantic vision. Such romanticism is fueled by a confusion between farming and agriculture.

Farming is a resource transformation process in which land, seed, and labor are converted into, for example, peanuts. It is Farmer Brown cultivating the land, sowing the seed, and harvesting the peanuts. Agriculture is the decision to invest money in this year's peanut production; the use of a tractor and cultivator to prepare the land; an automatic seeder for planting; application of herbicides, insecticides, fungicides, nematocides, and bacteriocides to kill unwanted pieces of the ecology; automatic harvest of the commodity; sale of the commodity to a processing company where it is ground up and emulsifiers, taste enhancers, stabilizers, and preservatives are added; packing in convenient, pleasing-to-the-consumer jars; and finally marketing under the trade name of "Skippy" or "Dippy." In short, while farming is the production of peanuts from the land, agriculture is the production of peanut butter from petroleum.[11] Over the last two hundred years, and especially in the last fifty, farming has become an ever smaller part of agriculture. This transformation has involved impressive technical as well as socioeconomic changes.

The Technical Side of Modern Agriculture

The transformation of farming to agriculture developed in two recognizable waves. The first wave commenced at the end of the Civil War and culminated in the mid-twentieth century with the emergence of the hydrocarbon society. The second wave started towards the end of World War II and is still evolving.

Before the outbreak of the U.S. Civil War, Cyrus McCormick built a revolutionary machine, a device you could attach to a team of horses and pull through the field to harvest automatically. Surprisingly, McCormick's automatic harvester did not come into use until well after the Civil War had begun, despite the fact that it was well known some twenty years earlier.

The all-important factor in the automatic reaper's initial neglect was availability of labor. At the time the U.S. Civil War broke out, the westward expansion of agriculture that had populated the Ohio Valley and was in the process of populating lands further west required a supply of harvest labor easily met by the comparatively large population of potential workers.[12] The many men (and very few women) looking to establish homesteads provided a ready source of cheap harvest labor. With all that cheap labor available, it made no sense to invest a large amount of money in an automatic harvester. Thus, McCormick's reaper, which eventually revolutionized agriculture in the United States, was ignored for two decades, not because farmers were unreasonable or backward, but because of rational economic calculations.

With the outbreak of the Civil War, labor grew short, and the McCormick reaper became widely adopted, initiating a change in the mentality of farmers. "Factories in the fields"[13] were to become the future of agriculture, and a host of automated devices, from corn harvesters to threshers to automatic seeders and more, followed McCormick's reaper. At this point, power was a limiting factor. The horse team was replaced by the steam engine for certain processes, but this was not always convenient.

The next great event within the first technological wave occurred in the early part of the twentieth century. The internal combustion engine made automated devices even more efficient. The introduction of the internal combustion engine—effectively the introduction of fossil fuel–based traction power—completed the first major transformation of *farming* to *agriculture.*

Finally, in conjunction with chemical research aimed at the development of weapons and the control of tropical diseases during World War II, a miracle seemed to emerge. The miracle was a chemical, dichlorodiphenyltrichloroethane, or DDT. Pests could be killed in the field much like enemies could be killed in war, and indeed the martial metaphor was skillfully used to promote the chemical transformation of American agriculture.[14] In fact, the munitions industry's burgeoning research into new explosives, which was based largely on nitrate chemistry, gave rise to technology for producing and using chemical fertilizers in agriculture.

U.S. agriculture soon became fully dependent on chemicals, pesticides, and fertilizers, most of which were formulated from petroleum. This completed the technological transformation of agriculture and set the foundations of "modern" agriculture. Today, a farm is a technologically sophisticated factory that uses fossil fuels to drive machinery that transforms the soil physically and chemically, uses products derived from fossil fuels to eliminate unwanted components of the ecology (i.e., pests), and adds other components that were either not there to start with or had become exhausted from prior use (i.e., nutrients).

It is necessary at this point to mention the "Green Revolution."[15] This term describes a set of policies, developed in the 1960s, whereby this modern agricultural system, mainly invented in the developed world, was exported to the Global South in the form of a technological package that included "improved" plant varieties and chemical pesticides and fertilizers. The claim of Green Revolution promoters was that the modern agricultural system would transform agricultural production in the Global South, saving millions from the ever-present danger of hunger resulting from overpopulation. Abundant evidence existed at the time, and still exists today, that rampant worldwide hunger results from poverty, not overpopulation. The ultimate goal of the Green Revolution was to create overseas markets for agricultural inputs (seeds, pesticides, fertilizers, etc.), not to fight hunger—and in fact, Green Revolution technologies have tended more to exacerbate hunger than to end it.[16] Today, the same agencies that promoted the original Green Revolution are promoting what they refer to as "Green Revolution II," which incorporates genetically modified crop varieties into the basic technological package. The critiques of the original Green Revolution apply equally to the newer version.[17]

The Socioeconomic Side of Modern Agriculture

In conjunction with this massive technological transformation, the socioeconomic structure of agricultural production underwent an equivalent major structural change. This transformation had to do with the way in which critical agricultural commodities were bought and sold. As far back in history as one cares to trace, it has been the norm that agricultural products were traded. Indeed, intermediaries were often involved. But at the close of the eighteenth century a food crisis loomed in western Europe. Wheat had always been a staple crop, but the form in which it was eaten was class dependent. The lower classes ate mainly porridge or gruel, souplike preparations that required little grain to fill a stomach. The upper classes ate a more elegant product, bread. To make bread, the wheat had to be ground into flour, mixed with water and yeast, kneaded into dough, allowed to rise, and finally baked. Such a large input of labor could be afforded only by the well-to-do. The peasant class continued to have to be satisfied with its porridge.[18]

During this period the long-distance trade in agricultural products was largely restricted to luxury goods such a chocolate, tea, coffee, and sugar. The main agricultural goods that fed the Industrial Revolution, cotton and wheat, were hardly traded internationally before the late eighteenth century. There was simply not enough demand since cloth was made mainly from wool and local production of wheat for porridge and bread fully satisfied all needs. But towards the end of the eighteenth century, two factors combined to dramatically change this pattern. The first factor was the evolution of improved technology to mill grain. The second was a change in the social organization of work.

With the growing industrialization of western Europe, the dominance of porridge and gruels began to decline, and the working class began to eat bread. Improved milling technology had caused the price of bread to decline, making it more available to the masses; more importantly, it was difficult for factory workers to bring gruel or porridge to work in their lunch buckets. The result was a dramatic increase in bread consumption, and a concomitant increase in the demand for wheat. In almost all European countries local agricultural production simply could not keep pace. This led to a dramatic increase in the international trade in wheat, which in turn led to the first important international wheat trading business complex the world had ever known. It included key points at Liverpool, Constantinople, and Odessa.

As part of the consolidation of the Russian Empire in the sixteenth century, colonists were given land in exchange for settling in southern Russia (what is today the Ukraine and Moldova). These colonists were given some of the finest agricultural real estate in the world, on par with the rich soils of the U.S. Great Plains, and the pampas of Argentina. Their product was sold at the most natural market, southern Russia and the entrance to the Black Sea, the town of Odessa. Odessa became the market for the hundreds of thousands of farms that dotted the surrounding countryside.

At that time the grain trade was a chaotic business. Grain was purchased directly, with little predictability about when, or how much, grain would be available to millers and bakers and thence the general populace. This was especially critical in industrializing countries, as the growing workforce was becoming increasingly dependent on bread. A shortage of this critical resource had the potential to create a great deal of social instability. Soon clever grain merchants, realizing that overland travel of a courier was far quicker than sea travel of a lumbering merchant ship, developed a system in which they took samples of grain from ships as they sailed through the narrow straits at Constantinople, carried them overland to the trading houses of Europe, and sold in-transit shipments based on samples. This was the beginning of "futures trading," a speculative financial strategy that would become extremely important to the industrial capitalist system.

Agreements to buy and sell grain shipments were usually made in clubs or informal gathering places such as coffeehouses. Liverpool became the center for buying grain destined for the mills of England. The wheeler-dealers of the time knew where to meet with one another. Gradually these centers became formalized and eventually evolved into the giant commodity trading centers we know so well today.

Thus, by the end of the eighteenth century in industrially burgeoning Europe, we had the beginnings of a complex grain trade which involved (1) a serf farmer who sold grain (or gave tribute) to (2) a landowner (usually nobility in Russia) who sold it to (3) a transporter who transported it to Odessa where (4) a shipper bought it and sailed to Constantinople where (5) a representative of the shipper took a sample and brought it overland to (6) a primitive commodity exchange center in Liverpool or another European city where (7) a local merchant bought it to be sold to (8) a miller. Into this system stepped several entrepreneurial types. A typical example was Leopold Louis-Dreyfus.

At the age of seventeen Louis-Dreyfus left his family's farm and sold his share of the season's wheat harvest. At the time, commerce in Russian wheat was in the doldrums due to the expansion of other international sources of wheat and to the dominance of Greek shipping interests in the Black Sea. Louis-Dreyfus spent several years in the stream of wandering merchants involved in the European grain trade. He amassed a considerable amount of capital in grain trading, established his own grain company, and took out large loans aimed at enlarging his enterprise. With his firm and finances in place, Louis-Dreyfus arrived in Odessa in the 1860s and purchased the bulk of the grain storage bins in the city. He then sent out his own agents, not to buy grain from farmers, but to sign contracts to purchase their grain in the future. At the other end of his enterprise, in western Europe, he sold those contracts for future delivery. By the 1870s he was contracting for wheat from the Russian hinterland, shipping by rail to his storage facilities in Odessa, shipping the grain on his freighters, and selling futures and grain to buyers in Hamburg, Bremen, Berlin, Mannheim, Duisburg, and Paris. The Louis-Dreyfus Company had become the first giant grain company in the world.

Similar stories can be told for four other giants, Bunge of Argentina, Continental of New York, Cargill of Minneapolis, and André of Switzerland. With Louis-Dreyfus, these five giants dominated the world grain trade much like the Seven Sisters[19] dominated the world petroleum trade. For all practical purposes, all the grain produced in the developed world must be sold to one of these giants. Since the first edition of this book, published in 1995, Continental sold its grain marketing division to Cargill (in 1999), and in 2001 the André company sold its Asian divisions to the Nobel Group and otherwise retired from the grain business. So the big five are now the big four (Cargill, Louis-Dreyfus, Bunge, and Nobel), although some analysts now include Archer Daniels Midland as a new giant in the business.

While the pattern we describe is most highly developed in the grain trade, similar structures developed in other sectors of the food industry with canned vegetables, processed foods, and so on. The story, beginning with the Industrial Revolution and continuing with only minor variation today, is a tendency for buyers of agricultural products to become large, and the tendency for most financial decisions to be made far from their source, the farmer. By the time Louis-Dreyfus arrived in Odessa, the serf farmers of Russia had little knowledge of, or control over, what happened to their grain after it left their farm. But those who purchased it were involved in high finance throughout the industrial world.

The Modern System in Summary

The above history sets the framework for understanding the modern agricultural system. On one hand, technological innovation has created a massive need for modern inputs into the agricultural system, which the farmer is forced to buy. On the other hand, socioeconomic "innovation" has created a massive marketing and processing structure to which the farmer must sell. The farmer is caught in the middle.

As a footnote to recent agricultural developments, the so-called enclave systems that made European colonialism possible still exist in certain corners of the world. Enclave production remains something of an anachronism in the modern world, and would not even be mentioned were it not so important, as we shall see, in areas experiencing rapid tropical rain forest deforestation.

5: THE MULTIPLE FACES OF AGRICULTURE IN THE MODERN WORLD SYSTEM

MUCH HAS BEEN WRITTEN about the emergence of the current world order, and as a product of the past five centuries of European expansion, the clear split between the first and third worlds, or what is today more commonly referred to as the Global North and the Global South. It is hardly debatable that the changes induced by this expansion were among the most spectacular in the history of humanity. The structure of today's world was established at the time European merchant societies took control of world trade from the Chinese/Arab monopoly. Europe used the vehicle of international trade and the incorporation of colonial outposts to supply raw materials to fuel their industrial development. For the most part these raw materials were either luxury products (e.g., sugar, spices); drugs (e.g., coffee, tea); or industrial inputs (e.g., wood, cotton, indigo). Most frequently they were agricultural products.

The Global South evolved from those early colonies, and industrial capitalism became the dominant form of production in the developed world. Yet the world came to operate as a single economic entity. The notion of industrial capitalism existing in the Global North and some more primitive form of capitalism existing in the Global South is outmoded. The Global South is not simply "waiting" for the sort of economic development that made the Global North what it is. Rather our globe is one world system, connected together in a complicated network. To understand how and why rain forests are disappearing, it is necessary to understand that network.

The Modern World System

The notion that the world is intricately connected, that it no longer makes sense to try understanding isolated pockets, such as nations, is most commonly associated with the sociologist Immanuel Wallerstein.[1] We agree with his general assessment, and add that isolated thematic pockets are similarly incomprehensible unless embedded in this global framework. For this reason, attempts at understanding tropical rain forest destruction in isolation have largely failed. As should be clear by now, the fate of the rain forest is intimately tied to various agricultural activities. These agricultural activities are embedded in larger structures, some retaining a connection to agriculture, some not.

In this chapter we outline the components of the modern system that impact on the phenomenon of deforestation. First, we consider the operation of developed world agriculture, and note its special form, distinct from that which dominated only a century ago. Second, we look more generally at the dynamics of developed world economies. Finally, we analyze Global South economics as related to developed world structures.

Agriculture in the Developed World Today[2]

As described in the previous chapter, historical developments gave rise to a particular model we refer to as the modern agricultural system. Modern agriculture has three units: the suppliers who supply the inputs to the farmer, the unit that is the farm itself (farmers), and the unit to which the farmer supplies the products. Immediately apparent are problems stemming from the very nature of this structural arrangement.

In Figure 5.1 we illustrate this relationship, showing the transfer of materials (Figure 5.1a), the transfer of money (Figure 5.1b, horizontal arrows), and the rational economic desires of all three components (Figure 5.1b, vertical arrows). The suppliers obviously wish to get the most money possible for their supplies, while the farmer wants to pay as little as possible for them. The buyers want to pay as little as possible for the commodities they purchase from the farmer, while the farmer wishes to get paid as much as possible. There are inevitable tensions here. If we really had that textbook fantasy, a world of perfect competition and free trade, one might expect that these tensions would resolve themselves to produce fair prices and terms of exchange among the units involved in this three-part process. But in the real world various forms of power alter that theoretical free market.

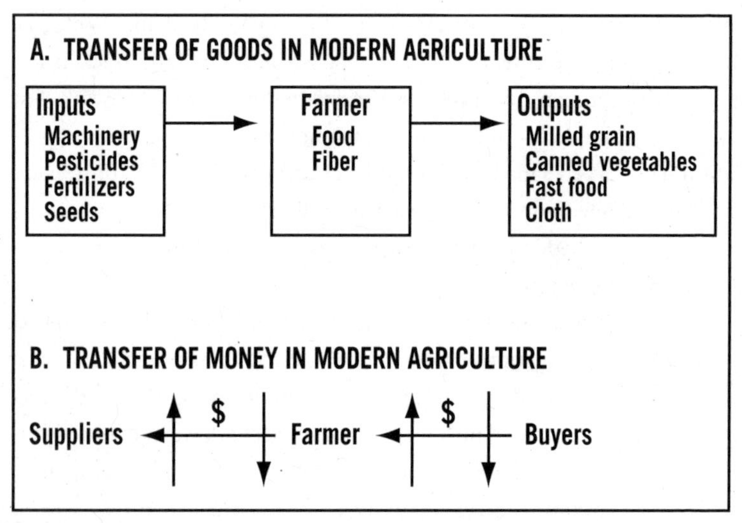

Figure 5.1. Diagram of goods (a) and money (b) transfers in the modern agricultural system.

The average U.S. farmer of the post–World War II era was a small entrepreneur running the so-called family farm. Such farmers had long before made the transition from only producing food for family consumption or to exchange for other goods to a more businesslike stance of investing money to make a profit. Yet the scale of operation remained exceedingly small when compared with suppliers like Dow Chemical or John Deere, giant corporations with monopoly-like control over the markets in which they participate, or equivalent sized buyers like the makers of Skippy peanut butter or the Cargill grain company. It does not take a Ph.D. in economics to predict what is likely to happen given such an arrangement. Farmers, with little economic leverage because of their isolated condition and relatively small size, were constantly squeezed by both ends of the system. Suppliers extorted large sums of money for necessary inputs, and buyers refused to pay fair prices for products (at least that is the way the farmer tends to see it), squeezing the poor farmer in the middle. This is the fundamental structure of agriculture in the U.S. and much of the rest of the developed world.

The implications of this modern structure are many and diverse, but for our purposes there is one critical feature. Agriculture has become part of the developed world's industrial system. While it may have made sense in the nineteenth century to speak of the "agricultural sector," and the "industrial sector," and to analyze them separately, such is not the case today. The independent farmer has become, in a sense, a worker in the metaphorical food factory that is modern agriculture. As such, developed-world agriculture today follows the same rules as the rest of the industrial system.

Developed World Economics[3]

Given today's interconnected world, in order to understand the Global South we must view it with reference to the operation of the Global North (the developed countries). We must view the Global South as embedded in the modern industrial system. In that system the people who provide the labor in the production processes are not the same people who provide the tools, machines, and factories. The former are the workers in the factories; the latter are the owners of the factories. The owners, who own the machines and tools, directly make the management decisions about all production processes. A good manager tries to minimize all production costs, including the cost of labor.

However, the owners of the factories face a complicated and contradictory task. While factory workers constitute a cost of production to be minimized, they also participate, along with the multitudes of other workers in society, in the consumption of products. In trying to maximize profits, factory owners are concerned that their factories' products sell for a high price. This can only happen if workers, in general, are making a lot of money. In contradistinction to what is desired at the level of the factory (for the workers to make as little as possible), at the level of society the opposite goal is sought (for the workers to make as much as possible). Factory owners must wear two hats then, one as owners of the factories, and another as members of a social class. As the owner, he or she wishes the laborers to receive as little as possible, but as a member of the social class, he or she benefits if laborers in general receive as much as possible (to enable them to purchase the products produced in the factory).

As factories compete with one another, labor, being one of the major costs of production, is always a target for reduction. Competition forces individual factories to mechanize processes whenever possible, giving them a competitive edge, at least temporarily, over those factories that have not mechanized. Unfortunately, mechanization overall means reduced employment, which means lowered general consumption power,[4] which means economic recession. But economic recession means a reserve of laborers ready to take up new employment at low wages, which stimulates the opening of new economic sectors and economic recovery. Since the new economic sectors comprise individual competing factories, the whole process starts over.

Phases of economic expansion and contraction are not desirable. Economic contraction is classically associated with political instability, something that factory owners and their political representatives naturally wish to avoid. Furthermore, and probably more importantly, economic contraction means profits are lowered and generally more difficult to come by. Lowered profits mean less money for investment and thus less chance of opening up new economic sectors to generate an economic revival. If new economic sectors do not appear, the general process of economic contraction will continue unabated.

Thus factory owners (as well as the general population) view such economic contraction as a crisis. On one hand, it is a crisis at a general level in that it can lead to the demise of the entire economic system. On the other hand, and more importantly for our purposes here, it is a crisis for individual factory owners in that their ultimate goal, to make a profit on investment, is threatened. Just how such crises are resolved is an important part of the history of the development of the industrial capitalist system, a point to which we return below.

Global South Economics and Dependency Theory

Difficult as life may get for some citizens of the United States or other developed nations, one can hardly fail to note a dramatic difference from conditions of life in the Global South—in the so-called developing nations. It is in the Global South that agricultural production has seemingly not been able to keep up with population growth. It is also in the Global South that dangerous production processes are located, where raw materials and labor are supplied to many industries at ridiculously low cost, where people live in desperation, and where talk of bettering the state of the environment is frequently met with astonishment: "How can you expect me to worry about deforestation when I must spend all my worry time on where I will find the next meal for my children and myself?"

Late nineteenth and early twentieth century analysts were generally in agreement on the basic mechanism causing such a pattern. It was thought that the Global South was simply behind, and given time, nations of the Global South would develop, just as the developed countries had. Consequently they became known as the "developing nations." This "time" hypothesis was the conventional wisdom for a long period. Yet by the end of World War II it had become evident that not only was the Global South not developing, the trend was actually the reverse. In most countries things were getting worse, not better.

This simple observation led to a great deal of analysis, beginning in the 1950s, but with the most vigorous discussion and debate occurring during the 1960s and 1970s. The general field of analysis is now known as dependency theory. In its most general form, dependency theory holds that the underdevelopment of the Global South is not an accident of history or a product of bad real estate, but an organic outgrowth of the progression of the developed world. Underdevelopment in the Global South is seen as an outgrowth of events in the Global North, and is somehow necessary for the maintenance of the Global North. Many dependency theorists have elaborated on this theme, and hundreds of books have been written, representing various positions within the general idea of dependency.[5] The one, and we think only, thing on which they agree, is that the underdevelopment of the Global South is a consequence of the development of the Global North.

To summarize the complex and historically conditioned debates within this vigorous field of political economy would require another entire book. Rather, let us examine what appears to be the most lucid explanation of the phenomenon, one that more or less summarizes what others have said and condenses the major arguments of dependency theorists into a series of simple ideas.[6]

Recall the characterization of the industrial system as consisting of two classes of people: the owners and the workers. The owners had to wear two hats in making decisions—wanting to pay their workers as little as possible, but wishing for workers in general to make as much as possible. This contradictory position represents the "engine" that drives economic growth in a modern capitalist economy. Production processes are mechanized to reduce labor costs, making more workers available at lower labor costs, so that new economic activities are then able to gain a foothold.

The situation in much of the Global South appears superficially similar. For the most part these are economies in which there are two obvious social classes: those who produce crops for export, like cotton, coffee, tea, rubber, bananas, chocolate, beef, and sugar, and those who produce food for their own consumption (or for a small local market) on their own small farms and, when necessary, provide the labor for export crop producers.

Consider, for example, Carlos, a Costa Rican farmer we used to know. He worked for the United Fruit Company for over ten years. Because of a temporary drop in the price of bananas on the international market, the company decided to scale back production. Carlos lost his job. United Fruit was the only employer in the area, so Carlos and his family were forced to either migrate to another area where a job might be found or find a piece of land and carve out a farm. They chose the latter.

The productive part of their farm is approximately two hectares, on which they grow cassava, taro, and several fruit trees. They also have chickens and a cow. Carlos feeds his family on the root crops, the eggs from his chickens, and the milk from his cow, and purchases some basic grains with the small amount of cassava he can sell. The last time we saw Carlos, a German company had started an ornamental plant farm up the road, and he had gotten a job there as a night watchman. He told us, "At least I won't forget what money looks like." His salary was about US$3 a day.

Another acquaintance, Bruce (not his real name), was chief executive in an international corporation involved in export crop production. After we sat in his living room listening to him close a million-dollar deal over the telephone, he recounted how his company was relocating much of its operation to Honduras where the military "knows how to control the unions," unlike in Costa Rica, where strong democratic traditions maintain the rights of workers to organize.[7] However, his main concern was with the threat of a U.S. recession, since his company exported luxury agricultural products there. That the Costa Rican unions would have made wages higher for Costa Rican workers and would eventually promote the growth of the economy was obviously of no concern to him (except of course in the negative sense that his company would have had to pay them more).

Superficially these people appear similar to the factory owner and worker of the Global North. Yes, Bruce runs a "factory" and Carlos works (or worked) in a "factory." But Bruce is not concerned with selling commodities from his production process to Carlos and people like him, but rather to consumers in the first world. Export producers of the Global South do not wear two hats with regard to the laborers on the farm. The concern is not to sell products to those workers, but rather to the workers in the Global North. What this means is that the dynamo of economic growth created by the contradictory goals of the factory owner in the Global North simply does not exist in the agrarian third world. The Global North factory owner is concerned with the general economic health of the Global North working class, and thus there is social pressure to maintain consumption power in that class. There is general agreement that while such an arrangement tends to produce economic cycles of boom and bust, it also represents the mechanism of economic growth, and is therefore the base of development. But to a significant extent that base of development does not exist in the third world. Bruce could care less whether Carlos can buy cotton or bananas or sugar. Bruce's concern is whether Detroit autoworkers and San Francisco yuppies can buy his products.

In Figure 5.2 this idea is represented diagrammatically, assuming the country in the Global North produces only automobiles and blue jeans and the country in the Global South produces cotton and bananas. Just follow the arrows (for example, the autoworkers and textile workers "pay money for cars to" the owner of the auto factory who "pays money for labor to" the autoworkers). We hope it is obvious from the diagram that in the developed world the money that goes to labor eventually goes to purchase the products, thus providing the machinery of economic growth. The Global South lacks an arrow connecting the export producers (factory owners) to the traditional farmers/workers in the process of consumption. The workers and factory owners of the developed world are the consumers of Global South products.

The situation in the Global North as pictured in Figure 5.2 may be described as *articulated.* Articulation is used here in the anatomical sense of having joints, or articulations, which are composed of connected segments. The typical arrangement in the developed world is an articulated economy, while that in the Global South is disarticulated, in that the two main sectors of the economy are not articulated—connected—with one another. Furthermore, since the production systems of the traditional agriculturalists and the export agriculturalists are not connected, the economy is also referred to as dual—the traditional sector operates relatively independently from the export sector.

This disarticulation, or dualism, goes a long way towards explaining the differences between analogous classes in the first and third worlds. Flower producers in Colombia do not concern themselves much over the fact that their workers cannot buy their products. On the other hand, the factory owners in the U.S., whether they be private factories or government owned and/or subsidized industries, care quite a lot that the working class has purchasing power. General Motors "cares" that the general population in the U.S. can afford to buy cars. Naturally they aim to pay their own workers as little as possible, but that goal is balanced by their wish for the workers in general to be good consumers.

So the different economic structures—articulated in the Global North, disarticulated or dual in the Global South—go a long way towards explaining the lack of development and the economic stagnation in those areas of the world. What remains to be explained, and this is a more difficult concept, is how this dualism in the Global South is maintained.

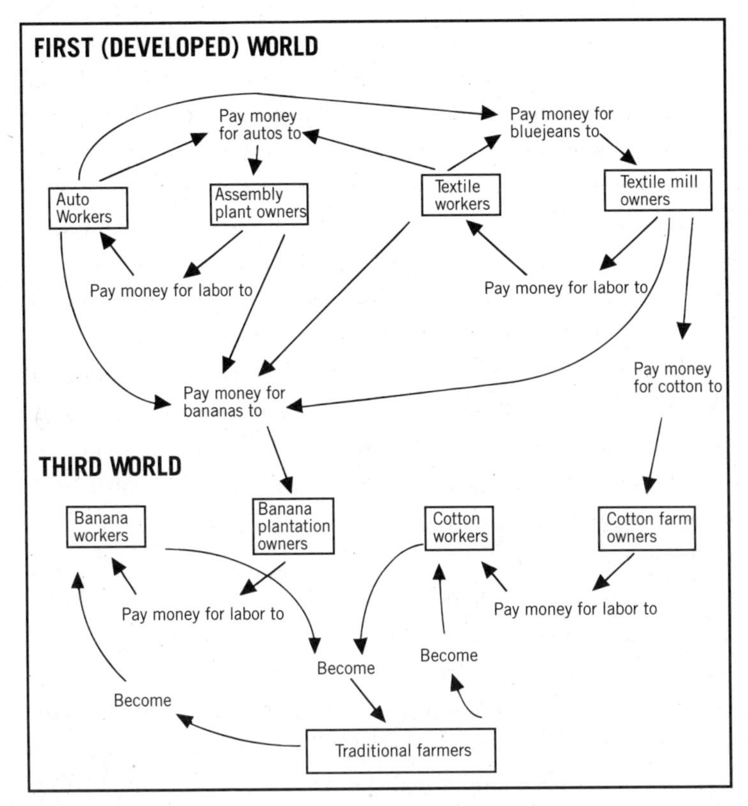

Figure 5.2. The world system.

The Function of the Dual Economy

The climate for investment is variable in the developed world. There are times when it is difficult to find profitable investments at home. At such times it is useful to have alternative investment opportunities. The Global South provides those opportunities. The developed world, because of its basic structure, tends to go through cycles of bust and boom, sometimes severe, other times merely annoying. During low economic times, where is an investor supposed to invest? Clearly without the presence of a third world, the crisis would be of a qualitatively different sort, in that the Global South provides a place for investments during rough times in the first world. This is why the dualism of the Global South is a "functional dualism." It functions to provide an escape valve for investors from the developed world. The German entrepreneur who started the ornamental plant farm on which Carlos worked as a night watchman invested his money in Costa Rica because opportunities in his native Germany were scarce at the time. What would he have done had there been no Carlos willing to work for practically nothing, and no Costa Rica willing to accept his investments at very low taxation? Clearly Costa Rica is, for him, a place to make his capital work until the situation clears up in Germany. Union Carbide located its infamous plant in Bhopal, India, and not in Grand Rapids, Michigan. U.S. pesticide companies export insecticides that have been banned at home to countries in the Global South. U.S. pharmaceutical companies pollute the ground water in Puerto Rico because they cannot do so (at least not so easily) in the fifty states. In all cases the people of the Global South are forced to accept such arrangements, largely because of their extreme underdeveloped economy.

With this analysis, it becomes clear that the origin of the Global South as an outgrowth of European expansion (while a correct and useful historical point) is not the only factor to consider when examining underdevelopment in the Global South. Even today, the maintenance of the underdeveloped world is a consequence of the way our world system operates. The developed world remains successful at economic development for two reasons: first, because it has an articulated economy; and second, because it is able to weather the storm of economic crisis by seeking investment opportunities in the underdeveloped world. The Global South in contrast has been so unsuccessful because its economy is disarticulated, lacking the connections that would make it grow in the same way as the developed world. Yet at a more macro scale, the dualism of the Global South is quite functional, in that it maintains the situation in which investors from the developed world can use the Global South as an escape valve in times of crisis. Indeed it appears that the developed world remains developed, at least in part, specifically because the underdeveloped world is underdeveloped. This is a major point of the dependency theorists.

This picture is very general and certainly does not strictly apply across the board. Many countries do not fit this model. For example, Taiwan and South Korea are Global South countries that experienced remarkable levels of economic growth; Hong Kong and Singapore have few peasant farmers; Finnish capitalists have few investments in the Global South; and none of the command economies in eastern Europe had significant investments in the third world. Each of these examples has particular exceptional circumstances associated with it. Far more common are the underdeveloped status of El Salvador, Nicaragua, Costa Rica, Colombia, the Philippines, Thailand, Vietnam, Mozambique, Zaire, Egypt, etc.

The Modern World System and Tropical Deforestation

Given such a socioeconomic structure, it is easy to see how tropical rain forests become targets for deforestation. At times of low market value for tropical export crops, rural proletarians lose their jobs and are forced to return to the countryside and a life of farming. Since most of the existing agricultural land is devoted to those very export crops, the former proletarian is forced to look elsewhere for land to farm. Frequently the rain forest is the only land available.

Complicating this picture is the position of the logger, as described in the next chapter. Cutting logs out of a tropical forest does not necessarily represent deforestation in and of itself. However, in the context of a disarticulated economy in the Global South, when logs are cut from a forest, landless peasants are nearly always waiting to follow the logging roads and take advantage of land that has been at least partially cleared of trees.

As we have already emphasized, lack of land and thus lack of food security is the driving force of deforestation. The logged forest could easily regenerate into a tropical rain forest were it not for the additional conversion to agriculture. So the force that creates landless peasants must be identified as a major force in the destruction of tropical rain forests. That force is the modern world capitalist system.

Yet there is a further complication to all of this. Export agriculture itself is also implicated in tropical rain forest destruction. Malaysian rain forests are directly cleared to make room for rubber plantations. Forests in Costa Rica are directly cleared by banana companies. While this direct form of conversion to modern agriculture is evident and even spectacular in several pockets (e.g., Brazil, Malaysia, Indonesia), it is probably not the most common form of rain forest destruction. When direct conversion occurs, it is probably more damaging than what peasant agriculturalists do, but its impact worldwide is less.[8]

However, the secondary impact of export agriculture is, as we have indicated in this chapter, dramatic. An additional component of this secondary impact is the need for Global South governments to continue expanding export agriculture. And this need is likewise an inevitable consequence of the underlying structure of the general world system.

We thus come to what we regard as a disturbing conclusion. An analysis of the socioeconomic aspects of current conditions in the world, especially as they relate to areas where most of the world's rain forests remain, suggests that the basic structure of that world system is what maintains the pattern of land and food insecurity that drives small-scale agriculture into rain forest areas. Saving the world's rain forests thus requires transforming that world system. A less disturbing, indeed optimistic, observation is that, since the first edition of this book, a new political movement has been born, one that has as its core a challenge to the basic structure of this world system. We describe that new movement in Chapter 7.

6: THE POLITICAL ECOLOGY OF LOGGING AND RELATED ACTIVITIES

CUTTING RAIN FOREST TREES is nothing new. The use of rain forest wood has been traditional for most human societies in contact with these ecosystems. But the European invasion of tropical lands accelerated the rate of wood cutting enormously, as tropical woods began contributing to the development of the modern industrial society.[1]

The direct consequences of tropical rain forest logging, apart from the obvious and frequently spectacular visual effects, are largely unknown. Some facts are deducible from general ecological principles, and a few studies have actually measured some of the consequences, but a detailed knowledge of the nontrivial direct consequences of logging is lacking.

What can be deduced from ecological principles is not that tropical forests are irreparably damaged by logging, but quite the contrary: tropical forests are potentially quite resilient. While this fact is to some extent debatable, most of the debate centers on how fast a forest will recover after a major disturbance such as logging (assuming nothing further occurs, like subsequent fires or agricultural activities), not on whether it will. The process of ecological succession inevitably begins after logging, and the proper question to ask then is how long it will take for the forest to recover.

In analyzing the effects of logging, we cannot assume a homogeneous process. There are a variety of logging techniques, some likely to lead to rapid forest recovery, others necessitating a longer period for recovery. For example, local residents frequently chop down trees for their own use as fence posts, for charcoal making, or for dugout canoes (Figure 6.1). Forest recovery after such an intrusion can be thought of as virtually instantaneous, since the removal of a single tree is similar to the creation of a light gap, a perfectly natural process that happens regularly in all forests. At the other extreme is clear-cutting—the extraction of all trees in an area. Though the physical nature of a clear-cut forest is spectacularly different from that of the mature forest, from other perspectives the damage is not quite so dramatic. The process of secondary succession, which begins immediately after such logging, leads rapidly to the establishment of secondary forest.[2] A great deal of biological diversity is contained in a secondary forest, and indeed, a late secondary forest is likely to appear indistinguishable from an old-growth forest to any but the most sophisticated observer, even though it may have grown out of a clear-cut. It may even be the case that large expanses of secondary forest contain more biological diversity than similar expanses of old-growth forest.[3] No studies thus far have been able to follow such an area to the point where it returns to a "mature" forest again.[4] One could estimate something on the order of forty to eighty years are needed, perhaps, before the forest begins to regain the structure of an old-growth forest.

Probably the most common type of commercial logging is not the clear-cutting described above, but rather selective logging. In an area of tropical forest that may contain 400 or more species of trees, only twenty or thirty will be of commercial importance.[5] Thus a logging company usually seeks out areas with particularly large concentrations of the valuable species, and ignores the rest. Often the wood is so valuable that it makes economic sense to build a road to extract just a few trees.

There are three negative consequences of selective logging that must be considered: road building, secondary damage, and deterioration of the stand. While direct damage to the forest from roads is considerable, it is the indirect consequences that have more important effects. Such roads offer access to the forest (Figure 6.2). Hunters, miners, and peasant agriculturalists have new and extensive access to areas difficult to get to before. In most situations this aspect of selective logging is probably the most con-

Figure 6.1. Informal uses of rain forest wood. Top photo is a dugout canoe being made from a downed *Ceiba pentandra* tree. Middle photo is of a completed dugout canoe. Bottom photo, a young Rama boy takes bags of charcoal to market in a dugout canoe. All photos form eastern Nigeria.

Figure 6.2. A recently logged area in southern Costa Rica. Note that there is forest on either side of the logging road. While many trunks had been removed, this particular operation left standing substantial amounts of vegetation, and the regrowth of the forest probably will be quite rapid. The road, on the other hand, will provide easy access for homesteading agriculturists.

ducive to deforestation.

"Secondary damage" from selective logging refers to the impact on nontarget trees during the process of removing the targeted ones (Figure 6.3). It is frequently extensive. Several studies have calculated that when a logging operation seeks only a small percentage of the trees in a forest, usually less than 10 percent, the incidental damage done by the machinery moving logs through the forest and by building roads results in many other trunks being knocked down.[6] However, we have observed small-scale operations with trunks hauled by oxen in which the damage is significantly lower. It would seem obvious that the techniques used in selective logging will partially determine the rate at which forests recuperate, which is dependent on the extent of initial damage.

A final aspect of selective logging, deterioration of the stand, must also be acknowledged. Removing only the valuable species and genetic types while leaving behind the less valuable ones results in an ever-decreasing fraction of trees in the recuperating stand that belong to the valuable

Figure 6.3. Bulldozer moves a recently felled log, scraping it along the forest floor, and in the process killing most if not all the advanced regeneration of the forest, probably setting back the regeneration time of the forest significantly. Photo taken in eastern Nicaragua.

species and genetic types. If a rotational system is initiated in which logs are cut every twenty or thirty years from the same site, it is easy to see how the stand may become less and less valuable as time goes by. In addition, if *all* the trees of a particular species are removed, that species can be driven to local extinction. This unfortunately occurs frequently with valuable species like mahogany. Furthermore, studies on the effects of selective logging on wildlife, which plays an important role in pollination and seed dispersal, have been contradictory and controversial.[7]

Reforestation and Tropical Plantations

Logging companies are frequently chastised because they are "cutting more forest than they are planting," and it is increasingly common for conservation-minded people from the north to promote brigades of tree planting volunteers to "restore" the rain forest. Such actions are sometimes misguided.

Rain forests contain countless species of plants and animals, and any attempt to restore them directly is absurd, even if all species were somehow available for restoration, which they usually are not. Furthermore, no one has anywhere near the requisite knowledge to advise how the restoration should proceed. How many tree species should be planted? Exactly where should they be planted? Should animals be introduced to disperse seeds? Should artificial perches be established to attract bird dispersers? Artificial bat roosts to attract bat dispersers? Should vines be planted or cut back? Given our current knowledge, calls for direct reforestation are at best naive.[8]

But the question of reforestation can be worse than simply naive. Contrast, for example, the "logged and left" area of the Danum Valley on the island of Borneo, in Sabah, Malaysia, with the "logged and reforested" area south of Kota Kinambalu, on the southwest side of Sabah. In the Danum Valley there is no attempt to engage in any reforestation activities and both selective and clear-cut logging have evidently been extensive. Vast areas have been converted from old-growth rain forest to secondary succession. Much of the area appears devastated, but a closer look reveals an area in the process of recovery. Pioneering tree species are beginning to form sections of diffuse canopy cover, and the process of ecological succession is well under way. The general area is currently a mosaic of various stages of forest regeneration, with old-growth forest near many rivers and streams, advanced second-growth forest in areas that were logged long ago, younger second-growth in more recently logged areas, and scattered areas of very retarded vegetation associated with logging roads and skidding areas. A biological survey of the area would likely find a majority of the species that had been in the area before logging, but their numbers and phenological stages would be significantly different. Trees that may have been common before are now present only as seedlings and saplings; insects of the deep understory are probably now restricted to the patches of old growth near the rivers; and the larger mammals have probably gone elsewhere, or are hiding in those old-growth patches. For the most part, however, the forest is actively regenerating, and the species that were there before are still there, albeit in reduced numbers. In short, the area appears more devastated than it actually is.

In contrast, the forest concession managed by Sabah Forest Industries has a different management plan. The old-growth forest is being cut down and replaced with plantations of a few fast-growing species, particularly *Acacia mangium.* The plantations form uniform stands of fast-growing trees, which will be harvested on an eight-year cycle to feed a giant pulp and paper mill. Does this represent "reforestation"? This is not a simple question and involves some tricky cost and benefit considerations.[9] But from the point of view of restoring a natural rain forest, the Sabah Forest Industries program appears to be far worse than nothing at all. A forest of hundreds of species of trees, the myriad insects that feed on them, and the innumerable organisms that live in the immeasurably diverse litter has been converted to a monocultural plantation. The loss of biological diversity has been almost complete. This is obviously by design, since the goal was to establish a plantation of *Acacia* trees.[10]

Frequently "reforestation" is precisely like the Sabah Forest Industries program. It is called reforestation, but because the goal is to create a uniform product to feed a sawmill or a pulp mill, species composition is dramatically reduced and then maintained in this restricted state. This is a plantation, and from the point of view of biological diversity, a plantation is more like a cattle pasture than a rain forest.[11] Furthermore, the fact that most of these fast-growing tree plantations are harvested in short (seven- to ten-year) cycles, brings up some questions about sustainability. Can a plantation like this be sustained over the long run? What will happen to the soil after the trees are cut and before the newly planted trees begin forming a protective canopy layer?

This brings us to the notoriously difficult issue of tropical tree plantations. Unlike temperate forests, which, because of their low species diversity, appear almost like plantations, tropical rain forests are dramatically more diverse. This diversity makes them both valuable and not valuable: valuable in that they contain scattered and rare high-value timber as well as many species of plants and animals that may have worth in the future; not valuable because the vast majority of the trees are of little commercial value. For this reason it is tempting, from an economic perspective, to turn the forest into one dominated by just the high-value trees. Thus far this has been a difficult proposition the world over. Mahogany, for example, has been planted in monocultures and usually fails because of a species of insect that attacks the growing stems.[12] The insects apparently have difficulty finding the scattered trees that occur in a natural rain forest, but travel easily from tree to tree in a plantation.

On the other hand, if tropical tree plantations can be made successful, they would provide much more valuable wood per hectare than natural forests. A successful tropical plantation would take considerable pressure off natural forests. Accordingly, it is a mistake to classify tropical plantations as useless. However, plantations must not be confused with natural forests. Creating a plantation is not the same as reforestation. Sometimes, perhaps most of the time, the attempt to reforest rain forest lands is really an attempt to establish tree plantations. This can take on pernicious political overtones. Several years ago in Nicaragua we were told a Swedish company was planning to establish large plantations of *Eucalyptus* and pines in a former rain forest area. This was in response to criticism from their home country's rain forest preservation movement that they were "cutting down forest and not reforesting." The confusion between creating a plantation and reforesting a rain forest allowed the company to continue logging rain forests while satisfying their political critics.

Another interesting example is the Ston Forestal company and its tree plantations in southern Costa Rica. Using the Asian *Melina* tree, Ston (now Smith Forestal) established 60,000 hectares of tree plantations,[13] sometimes cutting secondary forest in order to do so. The plan is to harvest the trees for chip wood and export the logs for home building and furniture making in the United States. Since the Costa Rican government offers tax credits for reforestation, the company insists that its efforts to establish plantations are actually "reforestation" efforts. Again, the confusion between reforestation and planting plantations has significant political consequences.

While there may be some exceptions, generally the best way to reforest a logged rain forest is to leave it alone. Rather than focusing attention on trying to actively restore a deforested area, energy would be better spent focusing on the nature of the logging operations themselves and on what happens after logging—for example, secondary agricultural incursions and forest fires.[14]

Disturbance and Recuperation in Rain Forests

A discussion of the political ecology of deforestation must occur within some basic ecological framework. As Blaikie and Brookfield[15] emphasize, notions like degradation (and by implication deforestation) must be carefully defined, for complicated subtleties in definitions suggested by different people and groups easily lead to confusion. To a botanist or an ecologist, a deforested area may mean any forest that has suffered from logging in the past, while from the point of view of a forester it may be an area with no harvestable timber. Or deforestation may refer to areas devoid of trees by social custom (e.g., cotton fields, cattle pastures). Coupled with this difficulty of definition is the growing acceptance among ecologists that disturbance is not generally a bad thing, but rather part of the normal ecology of a region.

The role of periodic disturbance has long been recognized in ecology as an important organizing force.[16] Certain ecosystems, such as prairies, savannas, and many forests, derive their structure from periodic fires. Tree falls create light gaps in a forest, initiating a process of succession on a micro scale, and possibly maintaining the high species diversity of tropical rain forests. Catastrophic wind damage is a well-known determinant of certain structural features of forests. Our work in the Atlantic lowlands of Nicaragua suggested direct and relatively rapid regeneration of the forest after a hurricane (Figure 6.4).

Whether the main form of disturbance is a hurricane or a natural tree fall, the implications of disturbance ecology are the same. It should be possible to design an extractive system that imitates natural disturbances in such a manner that recuperation of the forest is relatively rapid. Pickett[17] has suggested that the message of disturbance ecology for any sort of production activity might be to tune the production activities to mimic as much as possible normal disturbance events. The Palcazu project in Peru, where strips of forest are cut so as to mimic natural tree fall gaps, represents an implementation of this idea.[18] Perhaps foresters can gain some insights by studying the sort of damage done by a hurricane and how the

Figure 6.4. Natural disturbance is a recurring event in many lowland rain forests. Eastern Caribbean forest before disturbance (photo on top) and after passage of Hurricane Joan (photo on bottom).

forest recovers from it.[19] For example, a hurricane may topple or truncate a vast majority of the large trees, but do little damage to the understory, including the saplings that will eventually form the new tree layer. In a logging operation with heavy machinery, the understory is frequently damaged severely, sometimes virtually eliminated. The damage logging machinery does may be far more significant than the actual extraction of trees.

To some extent the outlines for new programs can be seen in the philosophies espoused (though not necessarily practiced) by some foresters, rather than in the one-dimensional programs commonly promoted by traditional conservationists. The summary of timber extraction in many tropical areas by Poore[20] and the excellent history of forestry by Westoby[21] are two of many possible primers. The point on which these and other forestry analyses agree is in the role of secondary activity in the logging operation—most importantly the invasion of agriculture into logged areas.

However, the story is yet more complicated, since agricultural conversion is not a homogeneous process. The nature of the agricultural activity is clearly an important variable. A small traditional peasant farm creates an environment that appears to share some features of tropical forest structure, with fruit trees forming an upper layer, other perennial crops forming a subcanopy layer, shade-tolerant plants in the understory, and patches here and there where the basic grains are grown in "light gaps."[22] When abandoned, such a farm is likely to revert to forest rapidly, provided there is a forest nearby to provide the necessary seeds. A peasant farmer who uses chemical pesticides may do further damage to the ecosystem, so that a return to forest after abandonment is somewhat more retarded. A small commercial farmer who uses machinery and chemical fertilizers and pesticides will do yet more damage, and a banana plantation or cattle ranch is likely to set back the process even further.

These observations lead us to the generalization that deforestation often proceeds as a three-stage process. Timber is extracted for commercial purposes, a disturbance whose recovery time is on the order of up to fifty years, and whose direct damage to biological diversity is only slight. Subsequent to timber extraction, the area may be converted to peasant or traditional agricultural activities, representing a disturbance whose recovery time is on the order of fifty to one hundred fifty years, and whose direct damage to biological diversity is significantly higher than the original logging operation. Subsequent to either traditional agriculture or timber extraction may be the introduction of intensive modern agricultural activity, representing a disturbance whose recovery time is on the order of one hundred to thousands of years, depending on the details of the disturbance. This general process is summarized in Table 6.1.

Table 6.1. The approximate damage to biodiversity and estimated time to recovery of various forms of human activities, based on general comments in the literature and our own experience. All figures are very rough guesses, as indicated in the text.

Activity	Damage to Biodiversity	Years to recovery
Informal wood harvesting	nil	0–5
Selective logging	nil to minor	10–20
Clear-cutting	minor	20–00
Traditional agriculture	minor to significant	50–150
"Modern" agriculture	dramatic	200–?

7: GLOBALIZATION AND THE NEW POLITICS

IN CHAPTERS 4 AND 5 we outlined how the international structure of agriculture relates to tropical deforestation, a pattern that is significantly more complicated today than it was ten years ago when we first wrote those two chapters. On one hand, there is a far greater consciousness about questions of tropical forest conservation and biodiversity issues in general among both political activists and the general public. Consequently, the job of convincing people that there is a problem is far less difficult than it was ten years ago.

On the other hand, there have been political changes on the global scale that provide for a great deal of optimism about the goals of rain forest preservation. The basic pattern that we described earlier was that a large corporation, like a banana company, extracted peasant farmers from their land, using them as workers until economic downturns and then leaving them to the whims of a labor market that was hardly capable of absorbing them, thus forcing them back to the land, which frequently meant moving into rain forest areas and clearing a bit of forest to make a farm for their family. That pattern has been somewhat modified in recent years, largely in response to international politics. Those international politics have two sides to them. First, in recent decades the major industrial capitalist countries have joined together to regulate world trade for their benefit through such organizations as the World Bank, the International Monetary Fund, and, especially, the World Trade Organization, and through "free" trade agreements such as the North American Free Trade Agreement (NAFTA). Second, and to some extent in response to the practices of the industrial countries, massive movements of workers, farmers, environmentalists, and other grassroots political movements have come together in unprecedented numbers and levels of cooperation to form what some refer to as the largest political movement in history. Aided by widespread communication technology such as the Internet, this movement gained worldwide attention in Seattle in 1999 and has been active and growing since. The largest political protest in the history of the world occurred before the U.S. invasion of Iraq in 2004. This certainly was not a direct protest against tropical deforestation, but it was an event made possible by this new and totally unique alignment of movements. And this new alignment significantly alters—we think for the better—the opportunities for taking action on tropical forest protection.

It is not the case that the new alignment grew out of nothing, and it is certainly not the case that it developed simply because of the Internet and cell phones. Rather, its precursors can be traced in many historical developments, especially since World War II, when the fundamental world system that we know today was set up. Understanding this historical trajectory is important if we wish to understand the movements of today.

Early Historical Background

World War II left Europe and Japan totally devastated. In addition to the toll in human lives and suffering, combatant countries spent more money on WWII than on all other previous wars put together. The U.S., however, emerged strong from the war. By 1947, the U.S. had accumulated 70 percent of the world's gold reserves. The UK had gone from being the world's greatest creditor to the world's greater debtor. Many countries had sold off most of their dollars and gold reserves, as well as their foreign investments, to pay for the war. Essentially, the war resulted in a massive transfer of money from Europe to the U.S. By 1945, Japan and the exhausted countries in Europe faced severe economic problems that frustrated reconstruction efforts—inflation, debt (mostly owed to the U.S), trade deficits, balance of payment deficits, and depleted gold and dollar supplies. The challenge of the Soviet Union and its allies was proclaimed, and it was palpable. In the face of this challenge, the last thing the West needed was an unstable world economy, which a devastated Europe and Japan could have created. Analysts across the globe agreed that the main factors that led to the war—including the Great Depression, the German malaise following World War I, the complete collapse of the German economy, and the expansionist desires of Japan—were mainly economic. In 1944, the Allies convened a conference at Bretton Woods, New Hampshire, to analyze all of these problems with an eye to averting them in the future. The importance of this conference cannot be overemphasized. Two critical institutions emerged: the World Bank, created to finance economic development, initially for the reconstruction of Europe; and the International Monetary Fund (IMF), assigned the task of maintaining world economic stability so as to avoid the devastating consequences of a world economic depression.

Realizing that much of the chaos that led to World War II was economic in origin, the policy makers at the Bretton Woods conference emphasized economic arrangements that would stabilize the world economic system. Central to the main goals was the idea of "exchange rate stabilization." Fixed exchange rates were established to stabilize currency fluctuations so that investors and traders would face less of a risk. All currencies were pegged to the U.S. dollar, and that was fixed at 1/35 of an ounce of gold. The basic ideas were seemingly sound if you looked at the system from a short-term point of view, but had serious drawbacks if viewed from a longer-term perspective. For example, in order to maintain the credibility of the fixed exchange rates, the actual exchange rate (the one people actually trade for on the street) could not be all that different from the theoretical fixed rate. To manage this need, central banks were forced to spend U.S. dollars to purchase their own currency (to create a bigger demand for their currency when its value began to fall on the street). Financial speculators could thus anticipate particular countries running out of U.S. dollars, which effectively placed great downward pressure on domestic money outside of the U.S. However, because the U.S. dollar was pegged to gold, there was no mechanism for the U.S. to lower the value of the U.S. dollar (which, of course, would have been another mechanism to maintain the credibility of foreign exchange rates).

To implement this foreign currency arrangement, the Bretton Woods conference created the International Monetary Fund (IMF) and the International Bank of Commerce, known generally as the World Bank. The IMF was assigned the task of ensuring economic stability, and thus preventing another global depression. So in its original conception the IMF was based on the recognition that markets often did not work well (i.e., that they could result in massive unemployment and might fail to make needed funds available to countries to help them restore their economies).[1] The three stated goals of the IMF were: (1) to promote international monetary cooperation; (2) to facilitate expansion of international trade; and (3) to promote exchange rate stability. The mechanisms at the IMF's disposal were basically two: (1) direct monetary bailouts to countries experiencing balance of payments deficits and (2) intervention in the structure of national economies—what was later to become known simply as *structural adjustment programs* or SAPs. Both functions of the IMF have come under severe criticism from many quarters. Critics have noted that SAPs in general have been devastating to many countries, mainly those countries in the South that began the postwar period in an economically depressed condition. The second, more structural, criticism of the IMF has been that its bailout programs have mainly benefited the international banking industry, forcing country after country to forgo development of social infrastructure (e.g., schools, health care facilities) to be able to repay international banks for the loans that were, sometimes, given out to military dictators or corrupt officials who used the money for personal gain. The critics of the IMF say, "Why bail out banks that have made bad loans?" while the defenders of the IMF say, "If the loans are not repaid, future loans will never be made." According to Chalmers Johnson,[2] the IMF has become "the premier instrument of deflation, as well as the most powerful unaccountable institution in the world. The IMF is essentially a covert arm of the U.S. Treasury, yet beyond congressional oversight because it is formally an international organization."

In addition to the IMF, the Bretton Woods conference set up the World Bank (formally titled the International Bank for Development). The World Bank was originally created for the reconstruction of Europe and

Japan, so devastated from World War II. But the bank rapidly evolved into the primary financier of development projects in the South, with a formal goal to reduce poverty worldwide. Projects are funded through grants and loans throughout the South with the intent of creating infrastructure that will lead to economic growth. Currently the countries of the Global South owe the World Bank more than 160 billion U.S. dollars. And, excluding China, poverty—measured by the ability of people to earn enough money to buy sufficient food to satisfy basic caloric requirements—is worse than when the World Bank was founded.

Today officials at the World Bank acknowledge failure at their most basic task of reducing poverty. However, the major criticisms of the World Bank most recently have been associated with the environmental and social problems created by its projects. For example, it has been necessary to resettle at least five million people as a consequence of the World Bank's program of building large dams. At least some of those people wind up as peasant farmers looking for new homesteads.

The Cold War and the Rebellions

In the wake of World War II, the tensions that became known as the "cold war" intensified dramatically. In 1949 the world's largest countries, China and India, declared themselves independent of European colonialism. While the direct and obvious effects were on England (and to a lesser extent Japan, which had attempted to take over England's role as imperial power in China), the worldwide effect was viewed with great anguish in the halls of power. While Gandhi provided the world with a major political tool (nonviolence), China was the most worrisome, since it experienced an openly Communist revolution and thus presented an enormous challenge for the cold warriors of the West. However, two subsequent events, both associated with control of natural resources, set the stage for the future. The first was in 1953 in Iran.

Iran, also a British colony, seemed to be following the lead of India and China declaring itself independent of colonialism, with the exception that its new president was elected by an openly democratic process. Mohammed Mossadegh, elected in 1951, was a popular figure whose tendencies and programs favored Iranian control over Iran's resources, especially petroleum. He was thus seen as a threat by both British and U.S. planners. Anglo-Persian (the oil company designated by Churchill to supply the British Navy with oil) had monopoly control over Iranian oil, and part of Mossadegh's political program was to regain national control over Iran's main natural resource. The British intelligence agency, MI6, effectively lost its assets in Iran with the election of Mossedegh and thus was unable to come to the aid of Anglo-Persian and the British government. MI6 called on the CIA to coordinate an effort to eliminate Mossadegh. The MI6-CIA plan called for the organization of street demonstrations against Mossadegh, teams of thugs to terrorize his supporters, and a propaganda campaign to discredit him. The campaign was successful, and in 1953 Mossadegh was overthrown and the Shah of Iran installed as Iran's leader, hand-picked by MI6 and the CIA.

With the success of Iran behind them, the CIA went to work in Guatemala, as we already described in Chapter 4. Jacobo Arbenz, the democratically elected president of Guatemala, was overthrown. What Arbenz had done was similar to what Mossadegh had done: he challenged the basic arrangement of a disarticulated economy (see Chapter 5) that supplied corporations in the first world with economic opportunity—in the form of petroleum for Anglo-Persian and of land for the United Fruit Company. Such a challenge seemed a harbinger of things to come, and caused a great deal of consternation in the halls of the politicians and their clients in the United States of America. If Guatemala or Iran, as examples of several dozen countries that performed similar functions, were not particularly important in and of themselves, what might happen in Costa Rica or Nicaragua or Honduras if Guatemala were to be successful? What would happen in the Middle East if Iran were to be successful? Indeed, shortly after Arbenz's victory and initial confrontations with the United Fruit Company, both Costa Rica and Honduras began making demands of the United Fruit Company, presumably buoyed by Arbenz's example.

The CIA victories in Iran and Guatemala gave the U.S. an exaggerated confidence, which led it directly into Vietnam. The Vietnamese Liberation Front had soundly defeated their former colonial masters, the French, and represented the same threats as Mossadegh in Iran and Arbenz in Guatemala. The CIA went to work, and eventually Vietnam was the site of a full military engagement. But this time, despite enormous and unrelenting military pressure from the U.S., the Vietnamese won the war, and the U.S. was never to be the same. It is worth reflecting on that war, in which 47,378 U.S. servicemen lost their lives—a horrendous loss of life for a cause that was suspect from the beginning. These 47,378 U.S. citizens were sacrificed because politicians feared the third world would seek control over its own destiny (going along with tradition, we parenthesize the facts that 1,323,748 Vietnamese military personnel lost their lives and that the number of civilians who died is effectively unknown).

Reflections on the political and moral aspects of this particular war would take up too much space. However, the Vietnam War ushered in several important changes in the world. First, there have been enormous political restrictions on the use of U.S. military forces ever since. It is even suggested that the senior President Bush was unable to "complete" the work of the Gulf War because he was politically unable to commit ground forces to invade Baghdad; and as we write these words, a great deal of political propaganda is being lavished on the U.S. population in an attempt to justify the junior Bush's occupation of Iraq. Such efforts would not have been necessary before Vietnam. Furthermore, Vietnam signaled an international delegitimization of the United States. Because of its central role in the defeat of Hitler in World War II, the U.S. had enormous moral authority after that war. That moral authority was dramatically squandered in Vietnam. Also, there was an attitude change in the Global South as a consequence of the war. The CIA's victories in Iran and Guatemala were more than cancelled in Vietnam, in that countries of the South regained a notion that they too could struggle for independence. The final and perhaps the most important consequence of the Vietnam War was that Richard Nixon was forced to eliminate the gold standard, which had stabilized the entire world monetary system since Bretton Woods. The cost of the war was so exorbitant that the only way to pay for it was by minting more money than the gold reserves would sustain, given the fixed standard of 1/35 of an ounce of gold for each dollar.

By eliminating the gold standard, the West was effectively returned to the monetary system of the nineteenth century, a notoriously unstable system. Indeed, the entire rationale for the Bretton Woods conference was to create economic stability in the world, and the principle vehicle was supposed to be the rock-solid dollar, which was rock solid because it was pegged to gold. That advance (if indeed it was an advance) was eliminated in 1974 when Nixon was forced into his action. The worldwide economic instability seen in recent years (during the 1990s and early 2000s, for example) is at least in part a consequence of that action (although some would argue that the original Bretton Woods institutions were flawed to begin with and doomed to failure anyway—but those arguments are beyond the scope of this book).

Iran, Guatemala, and Vietnam represent but three of the many foreign incursions in which the U.S. and its allies have been involved. Usually when these and similar events are brought to our attention, the purpose is to analyze the U.S. role and either criticize or laud its foreign policy and military adventures (or necessities, depending on your point of view). Our purpose here is somewhat different. We note that Mossadegh, Arbenz, and Ho Chi Min were members of the so-called third world who said, "We refuse to cooperate with a world system that is profoundly unfair." They said it in different ways and for different reasons (Iran wanted to control its own oil; Guatemala wanted to have sovereignty over its own land; Vietnam, over its sociopolitical future), but their general overriding reason was the same: refusal to cooperate in an overall world system characterized (as it was and still is) by northern rich countries and southern poor countries. They represent the early uprisings in the new global war—not the contrived, media-based wars against drugs or terrorism, but the war that aims to change the basic nature of the world system so that it functions for the benefit of all people, not just the rich in the North. We will return to this point after we discuss more fully what has been happening to that world system since the end of the cold war.

Globalization: Old Wine in New Bottles

Well before World War II, and even before the Industrial Revolution, there was a tendency, with improved technology, to move away from home ports and local production and into foreign lands. European entrance into the world economy, for example, was a project that lasted a couple of hundred years and involved the extensive use of slavery and the massive transfer of gold and silver from the Americas to India and China—indeed, Europe's entrance into world commerce was made possible only by precious metals from the Americas.[3] The process of "globalization" has been a continual one, especially during the past 500 years.

The most recent manifestation of the trend towards globalization is the augmentation of the Bretton Woods institutions with the newly created World Trade Organization (WTO), a successor to the older General Agreement on Tariffs and Trade (GATT). The overall goal of the WTO is to promote prosperity through international "free trade." It is attempting to do what was eventually done in Europe as part of the Industrial Revolution. That is, just as individual parishes in England had different economic rules and regulations and it was necessary to unify them all under the same set of economic regulations, the current corporate (frequently referred to as "neoliberal") globalization tendency embodied by the WTO is to unify economic rules at an international level. In this analogy, today's nation-state occupies the position held by the parish in preindustrial Britain.

The WTO is the organ that is intended to supply the overall supervision and legitimization of the new global order. Unfortunately, it currently has no accountability to anything resembling democratic institutions. It operates in secret and is not obligated to explain or justify any of its actions to the people in its member countries or to any other entity. Country representatives in the WTO are largely trade lawyers allied with large corporate interests, and they have thus far acted to smooth the way for corporate activities to be carried out without disruption wherever and whenever the corporations desire. The WTO, with its strict ideology of The Market as religion, has acted (and can be expected to act repeatedly in the future) to thwart local environmental and labor laws in favor of corporate interests.

Bananas Again

The very first case taken to the WTO by the United States, in 1995, was the case of Chiquita Brands, formerly the United Fruit Company—so infamous for its role in the overthrow of the Guatemalan government and subsequent terrorist acts in that country. The issue of Chiquita's WTO case was simple. As a historical legacy of European colonialism, many of the island nations of the eastern Caribbean (e.g., St. Lucia, Dominica, St. Vincent) have long received special treatment from the European countries that had formerly held them as colonies (mainly Britain and France). Many of these islands rely on bananas as their main export item, and their economic viability had long depended on the trade preferences they received from their historical "partners." Most bananas in these countries are produced by small farmers, whose ability to affect economic policy both at home and abroad is limited, unlike giants like Chiquita, which wield enormous political clout. For example, small farmers are unable to purchase fertilizers and other inputs at the subsidized prices that the big plantations enjoy, and their ability to avoid taxes within their own countries is limited by their weak political power base. Thus, their production "efficiency" is lower than that of the big plantations of Central and South America. They would clearly have trouble surviving if the trade preferences they received from European importers were eliminated. However, from the point of view of Chiquita, such a trade preference limited Chiquita's ability to penetrate some of the European markets. This became especially important with the fall of the Soviet Union and its eastern European partners in the early 1990s. Chiquita trade analysts expected a dramatic rise in demand for bananas from these newly opened markets. Chiquita CEO Carl Lindner and his staff presented their case to the Clinton administration, and Clinton dispatched his trade representative to the WTO to argue Chiquita's case.

In March of 1997 the WTO rendered a temporary judgment that only partially satisfied U.S. demands for free access to European markets. While waiting for the final WTO decision, the U.S. announced that it would impose tariffs on a host of European imports, ranging from handbags to greeting cards, sometimes with a 100 percent tariff. European negotiators pleaded with the U.S. to act within the rules established by the WTO, but the U.S. insisted it would do so only if the WTO ruled in its favor. In fact the WTO adjudicated that the U.S. trade sanctions themselves were in violation of WTO rules, given that they were being imposed unilaterally—illustrating the fact that you only have to play by the rules if you are weak. Finally, in July of 2001 the WTO came out with its final rulings, which involved complicated legal structures that effectively took the position advocated by the U.S. on behalf of Chiquita all along.

This episode is an interesting case for a variety of reasons. As was pointed out by many analysts at the time, there were no U.S. jobs at stake, since all the workers were in Latin America, and nothing about the case could have suggested that U.S. citizens would have been affected one way or another by either high or low tariffs in Europe. Nevertheless, the U.S. government took on the case extremely vigorously. Why? As is very well known by now, Chiquita's CEO, Carl Lindner, had been a major financial backer for both major political parties, and one does not have to be too much of a cynic to suggest that the Clinton administration was simply paying back what it owed. Expert political analysts argue that contributions to politicians and political parties are ideological and simply part of the democratic process, and are not for the purpose of purchasing influence. But they have never been able to explain this common occurrence of contributions to both parties and their candidates by very large corporations. Nonexperts clearly see such contributions as at best purchasing influence and at worst outright bribery. And as a final, remarkable twist, two reporters from the *Cincinnati Enquirer* were fired for an investigative report they did on Chiquita (Chiquita is based in Cincinnati). The report, devastating in its revelations of Chiquita's disgusting behavior in Central America ever since its founding, was never challenged regarding the facts it reported. The firings, and an explicit apology to Chiquita by the *Enquirer,* were based on the fact that some of the information for the report was obtained illegally, enabling Chiquita, with its extremely deep pockets, to threaten the *Enquirer* with legal action in the absence of an apology.

The Global Poor Fight Back

The newly globalized system, building on the operations of the IMF and World Bank, with the WTO as the new enforcer for the international bullies, seemed to be all in place and ready to create the new world order so gleefully proclaimed by the winners of the cold war. The fly in the ointment came in the 1999 meetings of the WTO in Seattle, Washington. A coalition of many progressive groups, including environmentalists and unionists, came out in force and effectively shut down the meeting. A great deal of world attention was focused on the WTO as a result of the fiasco in Seattle, and ever since the WTO has been unable to meet without a security apparatus that makes the whole operation look at best suspicious, at worst fascistic.

Many critics fear that inequality in the world is going to be worsened by the new capitalist globalization. Currently, the developed world contains about 17 percent of the world's population. Just looking at one measure, energy use, illustrates the current level of inequality between developed and developing nations: 17 percent of the people now use 70 percent of the energy resources of the world. That leaves about 30 percent of the available energy for the other 83 percent of the people. If we look at almost any other resource, the figures are similar. Some people suggest that leaving 83 percent of the world with only 30 percent of the world's goodies is immoral, some suggest it is unfair, and some suggest it is asking for trouble.

We can only imagine what must go on in the minds of the world's poorest people. Some are unaware of their situation. But when they do become aware, it appears that they act in a fairly predictable fashion. They say, "As 83 percent of the world's population, we deserve 83 percent of the world's resources, and if we can't get them, we will fight for them." Consciousness of great inequality has been the source of the world's revolutions in the past, and we see no reason to think that humans would act any differently now. Indeed, the only difference between today and 1776 or 1789 is that the relative condition of the have-nots is far worse today than it was in the past. The history of the world under European domination (i.e., about the last 500 years) is a story of wars and rumors of wars. Many of those wars, perhaps most, were about resources. The Vietnamese insisted that the people of Vietnam should be the owners of their resources, not the French or the Americans, and they fought for that principle. The Iranians insisted that the people of Iran should be the owners of their petroleum, not Anglo-Persian (or British Petroleum, Anglo-Persian's daughter company), and they fought for that principle. The Guatemalans insisted that the people of Guatemala should be the owners of their land, not the United Fruit Company, and they fought for that principle. Sometimes they lose, as they did in Guatemala. Sometimes they win, as they did in Vietnam. But they will always fight. And that unquenchable willingness to fight is the source of the new "anti-globalization" movement. What is different today is that leaders of peasant organizations in Guatemala are communicating with leaders of peasant organizations in Vietnam to coordinate their activities and their political directions. Great debates are happening, but not in the traditional isolation of the nation-state framework; rather they are taking place in the unifying air of international grassroots cooperation and solidarity.

As the poor farmers and workers of the Global South have been organizing themselves in one nation after another, many different political tendencies within the developed world have also come together to champion the same causes. Just ten years ago conventional wisdom posited a fundamental clash of interests between workers and environmentalists. Workers, it was argued, need jobs at any cost, and environmentalists want to save nature at the expense of those jobs. With this framing of the question, people were effectively asked to take the side of either labor or the environment. The past decade has seen a successful dismantling of this framework, and by 1999 in Seattle, environmentalists and labor leaders marched together demanding justice: one protest sign read "Teamsters and Turtles: Together at Last."

Finally, the tendency that gives all of this new political maneuvering its true strength is that the protesters of the developed world are coming together with the protesters of the Global South and are beginning to forge a new vision for the future. The World Social Forum, an international framework that brings together those opposed to corporate globalization, brought about 20,000 participants in Porto Alegre, Brazil, to its first gathering in 2001. In 2002 the forum attracted more than 50,000 participants, and by 2005 more than 150,000 representatives from almost every nation in the world united in their desire for a better world. It is a growing global movement that unites under the slogan "Another World Is Possible," and biodiversity and the preservation of tropical rain forests are two items at the top of its agenda. Some have called it the largest political movement in history and suggested that it cannot be stopped. We hope they are right.

8: RAIN FOREST CONSERVATION: THE DIRECT OR INDIRECT APPROACH?

IN 1990 when the World Resources Institute reported on deforestation rates for most countries in the world, some new and surprising data became available to environmentalists: Costa Rica claimed the dubious prize for first place.

For almost a decade, Costa Rica, partly because of its proximity to revolutionary Nicaragua, had received an enormous amount of development aid from the U.S., as part of the same overall strategy that brought war to Nicaragua.[1] Concomitant with this influx of U.S. dollars was the growth of a conservation movement, strongly influenced by North Americans, and with few parallels in the third world. Combining this conservation movement with an enormous amount of money, one would expect spectacular results. Indeed Costa Rica's conservation policies suggest that such expectations are justified: approximately 27 percent of its national territory lies under some sort of protected status; it has arguably the most impressive national park system in the third world (comprising 10 percent of its total land area[2]); and it boasts some of the most progressive forestry laws in the world.[3]

Conservation policies in Costa Rica were created on the assumption that rain forest destruction could be curbed by: (1) buying up and protecting large tracts of land; (2) passing legislation; (3) securing large sums of project money from foreigners; and (4) mounting a massive public relations campaign. Alas, these policies showed little concern about the impact of some crucial socioeconomic factors, including the fact that Costa Rica has one of the most uneven land distribution patterns in Central America.[4] The cherished belief of conservationists was that these policies were working. But the World Resources Institute figures showed otherwise. Despite Costa Rica's policies, during the 1980s the country's forests came down at an alarming rate of almost 7 percent per annum.[5] Such destruction paired with such impressive conservation policies certainly represents a reality check for all rain forest preservation advocates. Not only was deforestation higher than anyone had anticipated, it was the highest in the world!

Preserving the world's rain forest has become de rigueur in the developed world. But despite such readily available influences as the militant socialist agenda advocated by movement icon Chico Mendez,[6] the Western rain forest conservation movement has remained reluctant to link its work to social and political questions. While Westerners concerned about the rain forest should perhaps have been asking questions about the Indonesian military government and its transmigration program to Kalimantan, or the Guatemalan military's reported use of herbicides to clear vegetation in areas where guerrillas were suspected to be hiding, or Nicaragua's progressive land reform program, Costa Rica seemed to represent a far cleaner story. In Costa Rica, no military intervention, no socialism, no revolution, no anti-U.S. sentiment clouded the picture. Bird watching and nature appreciation tours were touted in the travel sections of all major U.S. papers. In addition, Costa Rican authorities were amenable to advocating all the goals and programs of the mainstream rain forest conservation movement. Consequently, the past thirty years have seen the truly remarkable growth of a conservation program in Costa Rica, but one whose aims and advocacies have remained almost totally isolated from some major sociopolitical questions plaguing Costa Rica. And this isolation is precisely the problem with the program.

There has been, we believe, a general failure in the mainstream rain forest conservation movement to address the underlying causes of rain forest destruction. Our purpose in writing this book—hopefully obvious by now—is to note and acknowledge those root causes. We are well aware that our analysis is at odds with much of the international community that seeks to preserve rain forests. We do not despair at this incongruity. The problem of rain forest destruction is far too important to leave to misguided proposals and ineffective solutions.[7] It is an unfortunate truth that "empire building," careerism, and even economic self-interest sometimes drive conservation programs and foster "analyses" that systematically exclude a search for the real root causes. Perhaps this is because these root causes create conditions in other spheres of life that the conservationists would not like to see challenged. Perhaps the same political arrangements that provide conservationists with the privilege to ponder such weighty questions as, for example, biodiversity, also create the impoverishment that forces peasants to cut down rain forests. If so, it would be prudent to keep those political arrangements out of the spotlight—for the mainstream conservationists, that is. But we, like the Lorax,[8] seek to speak for the forest. For that reason we intend to focus the spotlight on exactly those underlying causes.

In this chapter we draw attention to these political issues in the hopes of challenging what seems to have become a myopic and elitist view of rain forest destruction—a view often, unfortunately, tailored more to careers in conservation biology and the maintenance of good relations with potential funders than to a sincere desire to stop the progressive loss of the world's rain forests. We first revisit the Sarapiquí region of Costa Rica as described in Chapter 1, and detail some of the complexities that exist there. We then compare that site with a similar one in Nicaragua, elaborating patterns that existed there mainly during the 1980s. Finally, we add to the picture the remarkable complexities that emerged during the 1990s. Our intent is to demonstrate, first, that the mainstream rain forest preservationist view of the issue is hopelessly superficial and ineffective, and second, that an analysis of the entire web of socio-political-ecological forces is essential to understanding what in fact causes the destruction of rain forests.

The Sarapiquí Revisited

This book began with the example of the banana expansion in the Sarapiquí region of Costa Rica. It is fitting now that we return to that example (refer to the map in Figure 1.2) and describe some of the events of the late 1990s. As anticipated, the situation began deteriorating rapidly toward the end of the decade. The main players were an international tourism company, a well-connected conservation research organization, two local political groups, several absentee landlords, and some fruit companies (Figure 8.1).

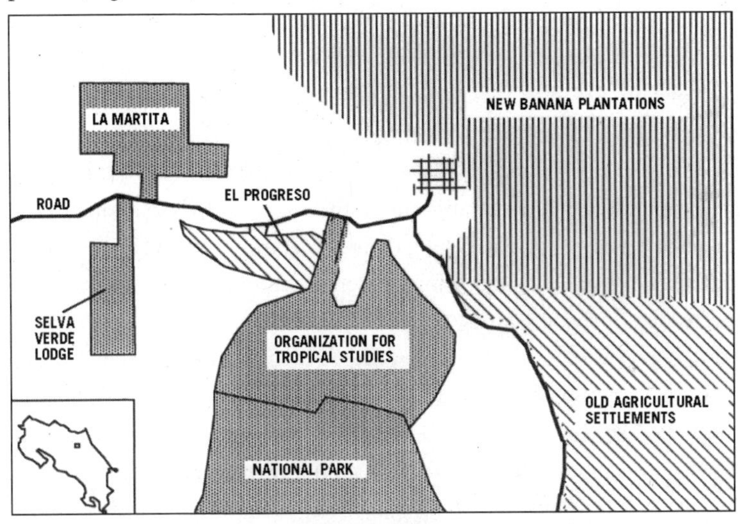

Figure 8.1. Sketch map of the region surrounding the town of Puerto Viejo in the Sarapiquí county of northern Costa Rica (insert shows location of area of detail), showing approximate locations of sites mentioned in the text.

The first local political group is a large, highly organized group of homesteaders in a community called "El Progreso." Their story began in the mid-1980s when a North American who owned a very large ranch in the area was indicted on drug trafficking charges[9] and fled the country. According to Costa Rican law, an absentee landlord can have his land expropriated if Costa Rican nationals establish a homestead thereon, which is exactly what happened. A Costa Rican couple who had worked for the North American established themselves as homesteaders and petitioned the appropriate government agency for title to the land. Meanwhile the North American got wind of what was happening and solicited the aid of a lawyer friend in Costa Rica. The lawyer reportedly filed papers aimed at incorporating the ranch into a group of farms he already owned.

In this way the farm, which the Costa Rican couple had renamed

"Gerika," came to have two or three parties claiming ownership: the indicted North American, his lawyer friend, and the Costa Rican couple. Then in August of 1993 tragedy struck the Costa Rican couple: the wife murdered the husband and then killed herself. In such circumstances Costa Rican law and tradition say that if there is a known long-term caretaker of the farm, ownership passes on to him or her. In this case a woman had been the caretaker, and she quickly laid claim to the farm. But another Costa Rican law allows for homesteaders to occupy unused land. With the bizarre death of the Costa Rican couple, who most locals had assumed were the owners, homesteaders saw the opportunity to seize land that had been unused for quite some time. Only weeks after the couple had passed away, homesteaders took over part of Gerika, calling their new community "El Progreso."

The takeover was not without violence, and a combination of angry homesteaders, private guards, and Costa Rican security forces produced some ugly incidents, including one death.[10] But a year after the establishment of the homesteading site, El Progreso was home to 350 families who had obtained legal "right to possession," the first step towards getting official title to the land under the Costa Rican homesteading act.

Several interesting facts highlight the El Progreso affair. First, some unknown number of the homesteading families were Nicaraguan (local estimates at that time ranged from "not many" to "almost all"), most of whom were attracted to the area by the promise of employment in local banana companies in the early 1990s. Second, the back border of the farmable land claimed by the homesteaders abuts directly on property owned by the Organization for Tropical Studies, one of the major conservation players in the area. Third, many in El Progreso are currently employed by the banana companies, which send buses to the community each morning to pick them up, and many others are ex-employees of the banana companies. Finally, evidently the real estate the El Progreso settlers have occupied is fairly rotten. While the members of the community are optimistic about their dreamed-of farms, we doubt much will come of their agricultural activities. The soils seem to be old alluvium, but are probably very acid and lacking in nutrients.

The second major local political group is the Sarapiquí Association for Forests and Wildlife. Their goal is to "promote the conservation and sustainable development of the natural and human resources in the county of Sarapiquí."[11] The association was formed by local residents with the help of a nongovernmental organization that provides legal assistance to local grassroots groups seeking to develop conservation projects.[12] The Sarapiquí Association's immediate objective was to purchase a piece of land called "La Martita," whose Colombian owner had applied for a logging permit from the government. As the association points out, several biological reserves and stands of old-growth forest exist in the Sarapiquí, but all are owned by foreigners. With the development of the ecotourism industry, local residents garnered some benefit as employees of ecotourism entrepreneurs, but there seemed to be no example of a local resident who was owner of a successful ecotourism establishment. The goal of the Sarapiquí Association was to use La Martita as a community biological preserve where locals could begin developing their own ecotourism operations (see Figure 8.1).

Much of the Sarapiquí was still owned by absentee landlords at this time, both Costa Rican and foreign. Excluding the banana companies, the main activity on these lands is cattle ranching, although many still contain substantial tracts of rain forest. Cattle productivity is poor in the region, and it has been suggested that the only reason they are kept on the land is to claim that the land is in "use" so it is not subject to invasion by homesteaders. La Martita is one such operation. The part of the property that connects with the main road is in cattle pasture and boasts a sign in front that reads "Warning, vicious dogs." The back of the property, the forested part, is not accessible from the main road, so it has not yet been a target of homesteaders. However, it is not out of the question to envision another group of ex-banana workers establishing a homesteading community here if the opportunity arises. If the association were successful in purchasing the land and preserving the forest, such an invasion would be far less likely.

The international tourism company is Holbrook Travel Inc., and they own the Selva Verde Lodge (see Figure 8.1). Styled after the highly successful nature tourism industry of Kenya, this lodge has upscale facilities that shield the tourists from the local people, except for those who work as cooks, maids, or maintenance workers. This impressive facility includes one of the largest privately owned nature reserves in the area, and is among the largest employers in the local ecotourism business. Local people most frequently cite the Selva Verde Lodge when they speak about foreign ownership of the local ecotourism business.

The other large ecotourism operation is the Organization for Tropical Studies (OTS). OTS is a consortium of universities in the U.S. and Latin America, with some fifty members. Its goals are research and education in tropical biology, and it is one of the most famous organizations of its kind in the world. Because OTS attracts rich foreign ecotourists and is owned mainly by non–Costa Ricans, the organization has come to be viewed by many local people as similar to Selva Verde. Such a perception is natural.

Finally, and still the predominant feature of the area, are the banana companies, dominated by the multinationals Chiquita, Dole, and Geest. As we predicted in Chapter 1, by the late 1990s the expansive phase of their operation had ceased, and many workers had been laid off. These workers and their families wound up living in homesteading communities like El Progreso, homesteading individually in various corners of the landscape, seeking employment wherever they could get it (there's not much work in the area other than the banana plantations), and increasingly migrating to San José, the capital. The situation of the workers got so bad that the all but defunct banana unions reappeared on the scene.

This then was the landscape towards the end of the 1990s. Ecotourism and international conservation groups rapidly losing credibility among the local population; banana companies attracting workers to the region and dumping them onto the landscape; absentee ranchers hiring armed thugs to protect their land from homesteaders; international tourists who could spot the rarest of birds through their binoculars yet could not see the desperate and hungry people just across the road; 350 hopeful and optimistic families in a new community with a utopian plan to turn acid soils into productive agriculture; and a weak, largely ignored local conservation group trying to get something for Costa Ricans out of ecotourism.

That landscape would seem to leave little reason for optimism. But we believe that this is mostly because analysis of the problem has been made in a fragmented fashion by the various interest groups. Local conservationists are concerned with preserving whatever forest remains, and see the homesteaders as conniving ne'er-do-wells. Some conservationists even attribute rain forest destruction to overpopulation. Homesteaders see large tracts of land unused by foreign owners and view their own landless situation as unfair. Gaining property title to a piece of logged-over forest seems logical and fair to them. The central government continues its policy of promoting the expansion of bananas, even while promoting conservation programs in other areas of the country. And lastly, the banana companies see the conservationists as romantic idealists who simply do not understand the nature of business, which is to make profits, nor the main need of the country, which is to generate revenues to pay off its international debt.

If, however, the problem is viewed in all of its complexity, and a plan devised that takes all parties into consideration, there is much that can be done. We might envision, tentatively, a three-part program. First, local people need jobs. We regard the question of food security as the underlying basis of any sound conservation policy. To this end, the labor practices of the banana companies must be regulated. No longer should they be allowed to hire workers for eighty-nine days and then fire them because the companies become liable for various social security benefits after ninety days. Working conditions must be improved, and above all job security must be instituted. While there is little that can be done about the banana plantations already in existence, a moratorium should be immediately implemented concerning the further expansion of such production technology.

However, there are clearly not enough jobs in the banana industry to employ everyone in the region. It should be noted that during the banana expansion of the early 1990s the banana companies asserted there were not enough workers in the region and, in coordination with the Costa Rican government, brought in Nicaraguan workers. Rumors of available jobs attracted people from all over Costa Rica and Nicaragua, dramatically increasing the region's population. The only obvious beneficiary of this chaotic migration process has been the banana companies, which now enjoy a surplus of workers.

The reality remains that not everyone will get a job in the banana industry. Thus, the second part of the program should support existing homesteader communities, beginning with El Progreso, with technical aid and bank credit to make their farming operations as viable as possible. It should be noted that such viability may not, in fact, be technically feasible: despite twenty-five years of active biological inquiry in the area, only a tiny fraction of that research is even theoretically relevant to finding solutions to the many problems faced by farmers in the region. And if the current agricultural areas of El Progreso turn out to be difficult to farm on a continuous basis, El Progreso's farmers may expand their agricultural activities into the forested areas that border their farms. It would seem that making El Progreso's agriculture successful should be an obvious priority, even for the ecotourism ventures.

Third, local people must directly benefit from the increase in ecotourism in the region, not just as janitors, cooks, and badly paid tour guides, but as entrepreneurs and owners of lodges and reserves. One part of accomplishing this would be support for the Sarapiquí Association for Forests and Wildlife, especially with its efforts to purchase La Martita and turn it into a community-owned forest reserve. Part of this general goal could easily be coupled with the agricultural imperatives for the region. Ecotourists are often interested in seeing alternatives to destructive agriculture and the people of El Progreso already have an impressive conservation ethic. Several homesteaders independently offered the opinion that deforestation was the country's biggest problem right now, and that ecotourism should be promoted. They see potential jobs for themselves in ecotourism. And, if they begin an experimental program of sustainable agriculture and agroforestry, their land might produce more than just corn and beans; it might produce tourist dollars.

What chance might this three-part program have for success? Given the constraints of the real world, probably very close to zero. But each component is worth pursuing. The alternative—foreign conservation organizations and ecotourism companies dominating all the forested land, banana companies dominating the remainder and hiring and firing workers chaotically—will certainly not work in the long run. Poor people need to feed themselves and their families. They need steady jobs or access to good agricultural land and technology. If they have no job, and no hope of getting one, if they are excluded from the promise of the bright future they can see the foreigners already have, if their crops continue to fail on the small piece of rotten land they have grudgingly been allocated, they will do whatever they deem necessary to survive. Even if that means cutting and burning primary forest on preserved land.

The RAAS of Nicaragua: A Contrast[13]

The story in Nicaragua is quite different from that in Costa Rica, although it also illustrates features of the general pattern we have described. The Atlantic coast of Nicaragua contains the largest remaining tract of lowland rain forest in Central America,[14] with significant closed canopy forest, or at least patches of it, extending from the border of Honduras all the way south to the border of Costa Rica (see map in Figure 1.2). There are, however, several foci of agricultural expansion. For the past few decades the largest of these agricultural frontiers has diffused eastward toward the area surrounding Bluefields on the Atlantic coast. Since 1990, the rate of expansion has increased dramatically, perhaps aided by the devastating effects of the hurricane of 1988 and the elections of 1989. The agricultural frontier has effectively cut the Atlantic coast rain forest into a northern and southern section, and is now expanding both northward and southward (see Figure 1.2). If this pattern continues we can expect the destruction of all but small islands of protected forest, with little potential for stable economic activity for the local population.

The majority of the upland area surrounding Bluefields (about 400,000 hectares) was originally covered with lowland tropical rain forest, almost all of which was severely damaged in 1988 by Hurricane Joan. The forest is now in various stages of ecological succession. Despite the severe destruction, our studies indicate that natural resprouting and direct regeneration have resulted in a habitat well on the way to mending itself. Small peasant farms dot the area and continue expansion into formerly forested land. A large sugar plantation is located north of Bluefields, and an old, mainly abandoned, extensive cattle operation covers an upland area known as Loma de Mico. For the past decade the sugar operation has rarely been profitable, and the cattle operation was virtually abandoned in the mid-1980s. Finally, a large African oil palm plantation was planted in the mid-1980s and produced at approximately 25 percent productive capacity until about 2001, when it ceased production.

The human cultural history of the Atlantic coastal area is extremely complex. The bulk of the population is about evenly divided between African descendants brought from Jamaica as slave laborers in the 1800s and mestizos who migrated from the Pacific coast during the past two centuries. Additionally, three small Native American groups inhabit the area: the Ramas; the Sumos, who have small communities north of Bluefields; and the Miskitos, most of whom live in the North Atlantic zone, with small numbers in the South Atlantic zone, living in either small villages or as individuals among the other cultural groups. Another ethnic group, the Garifuna, originally derives from escaped African slaves. While all groups maintain strong cultural identities, they are also united in a coast culture and regard themselves as *costeños* (coast people).

The political panorama is as complex as the cultural. The well-known friction between the Sandinista party and the former UNO (United Nicaraguan Opposition) is complicated by the presence of YATAMA (another political party representing mainly the Miskitos), and the entire political framework is permeated with a debate over autonomy. Historically isolated from the Nicaraguan mainstream, the Atlantic coastal region has always been fiercely independent in politics and economics. In 1989 the first autonomous administration was elected, and—at least on paper—the RAAS (South Atlantic Autonomous Region) is supposed to be independent of central government administration. This formal independence extends to decisions on all natural resources, including rain forests.

The RAAS encompasses the southern half of the Atlantic coast of Nicaragua (see map in Figure 1.2). Economically, it is something of a disaster. While Nicaragua is second only to Haiti in poverty in this hemisphere,[15] the Atlantic coast is the most underdeveloped region of Nicaragua. Local government officials claim an unemployment rate of over 90 percent in Bluefields, and a short visit to the area convinces even a casual observer that this is indeed the poorest region of Central America.

During the years of the Somoza dictatorship (1930 to 1979), the area was relatively isolated, and while Nicaragua as a whole is reported to have experienced massive deforestation during that time, little evidence exists for such deforestation in the RAAS in particular. During those decades, the agricultural frontier slowly expanded towards Bluefields, the large sugar plantation and cattle operation opened up, and timber extraction, primarily by North American companies, was more or less continuous. Though it is not known for certain, it is probable that, as elaborated above, the local logging operations were followed by peasant agriculture.

Following the Sandinista Revolution of 1979, two major developments relevant to tropical forest use came into play. First, the contra war, which was organized by the CIA, devastated the entire country. The war made it especially hard to engage in large forestry operations; it also made it difficult, in areas constantly threatened by fighting, for peasants to homestead or create new agricultural plots. However, casual interviews with people in Bluefields do not suggest that large numbers of landless peasants were champing at the bit waiting for the war to end so they could carve out a homestead from the remaining forest. To the contrary, the landless peasantry seems to have largely disappeared due to the second development: the Sandinista administration's agrarian reform program. With most rural families gaining title to land, the underlying push to move on and carve a homestead out of the wilderness was largely eliminated. Consequently, the rate of deforestation in the region was relatively low during the eighties.[16]

The elections of 1990, in which the Sandinistas lost power, changed everything. The progressive agrarian reform of the Sandinistas was reversed, and once again landless peasants were forced to seek new homesteads, frequently after having been forcibly removed from land acquired under the Sandinista reforms. As part of a deal with the new government, so-called poles of development (small areas designed for settling ex-combatants) formed new bases of agricultural expansion. Former contra rebels were given forested land to distribute among their "soldiers" and their families to establish agricultural communities. A typical pattern of land use following this type of redistribution consists of two or three years of cultivation after cutting a piece of forest, followed by a ten-year fallow period, followed by a single year of cultivation. This pattern understandably engenders in the farmer a strong desire to move on in hopes of finding a better piece of real estate under the next parcel of forest.

The relative stability the Sandinista agrarian reform provided now seems to have been destroyed. We have been traveling in this area since 1982, and the changes since 1990 have been frightening. Before the 1990 elections there was a great deal of debate on such questions as use of the natural resources of the area, the role of the state in regulating chain saws, how the state logging company (the only logging company operating) should be responsible to local communities, and how agrarian reform must extend into the technical sphere to provide peasants an alternative to burning for land preparation. By 1993, the constant drone of chain saws could be heard anywhere there was forest. Three lumber companies were actively engaged in cutting trees (openly flouting a ban on cutting), and central government members were accepting bribes for allowing logging concessions.[17] Fires were everywhere as landless peasants burned yet another piece of forest, and former contra bands, apparently frustrated with their inability to produce anything on the poor soils of the region, engaged in robbing passing boats and fighting with one another.

In addition, the economic situation has deteriorated significantly. The peasantry, which in the past could hope that the government would help in times of crisis (e.g., if the crops failed there was reasonable expectation that the government would supply the community with basic grains), has become despondent, as they are often forced to do anything possible to eke out a living, usually with little success. A typical story was related to us by the residents of a peasant community on the Patchy River, north of Bluefields. The community, a group of ex-contra rebels, had been resettled in the area under the leadership of their former commander, whose nom de guerre was Ranger. Ranger arranged with outside contractors to sell the mahogany and rosewood from the forest behind the community's small settlement. The community had neither beasts of burden nor tractors, so Ranger paid the peasants the equivalent of $3.50 per trunk (each of which would retail for at least $1,000 on the world market) to haul the logs down to the river dock by hand. In the end, Ranger disappeared with the logs and didn't even pay the $3.50 per trunk he had promised.

The Complications of the 2000s

The 2000s have not brought much positive news regarding rain forests. Nicaragua, suffering from extreme poverty, continues to view its rain forests as an escape valve for social problems that might otherwise cause those with power some consternation. Costa Rica continues its long tradition of love-hate relationships with the banana companies, with some deviation from the pattern we reported earlier. Although the rate of deforestation reported for Costa Rica for the period 1990 to 2000 is significantly lower than for the previous decade (0.8 percent), this is partially due to the high rate of establishment of monocultural plantations. By 2002 Costa Rica had 178,000 hectares of these plantations. Nicaraguans, facing a continually crumbling economy, proceed with their migrations to Costa Rica seeking work in the banana plantations, and then finding themselves out of work with no prospects other than homesteading. And much of the rest of the tropics seem mired in the neoliberal economic model, which encourages the same sort of social relations that were at the base of deforestation in the 1980s and 1990s. Anticipating this new edition of *Breakfast of Biodiversity*, we had hoped to report on some positive progress. Only by grasping at straws would we be able to do so.

Caribbean Coast Tensions in Nicaragua

Nicaragua veered to the political right in 1990 and even further to the right during the elections of 1996. The Constitutional Liberal Party (PLC) had a massive win in the national elections of 1996, leaving the Sandinistas a minority in the national legislature. The president elected in 1996 was Arnoldo Aleman, a long-time foe of the Sandinistas who seems to have a personal vendetta against them. He is a strong advocate of the neoliberal model in its most pure form, promoting the privatization of everything. Furthermore, many of the large landlords who left the country during the Sandinista administration—and whose land was thus confiscated and redistributed to the peasants—have been returning to reclaim their land. The peasants who held title to that land under the Sandinistas' agrarian reform program have been effectively thrown off of the land, although in most cases with compensation (there is some debate as to whether the compensation is just). Consequently, we have seen over the past decade an effective reversal of the Sandinistas' agrarian reform program. Since that program was effective in reducing pressure on tropical rain forests, we might expect renewed pressure resulting from the new agrarian structure.

Indeed, that is what we find.

Aleman was replaced by Enrique Bolaños, a member of the same political party, in 2002. Ironically, Aleman, who originally ran on a platform that promised reform of the corrupt bureaucracy the PLC had put in place, was himself indicted on corruption charges and eventually convicted and sentenced to twenty years in prison. The Sandinistas, now a more traditional political party, collaborated with Bolaños in the conviction and eventually connived with the PLC to create election laws that effectively created a monopoly for the two main political parties. This suggested to some analysts that the Sandinistas' long-standing claim to revolutionary reform had been replaced by the exigent behavior of a typical "modern" neoliberal political party.

The tropical rain forests of Nicaragua are concentrated on the Atlantic coast of the country, where a multiethnic mix of political factions are in continual contention over natural resource issues, including forests and agriculture. Miskito Indians, Black Creoles, and mixed-blood mestizos are the three dominant ethnic groups there, and the contradictory nature of their views towards managing the resources of the region have recently intensified as a result of the national trend towards the neoliberal model. Miskitos have large expanses of land they have traditionally used for hunting and the extraction of timber and nontimber forest products, which is to say they have been preserving the rain forest as part of their traditional culture. Creoles are more than anything focused on marine resources, but also engage in traditional agricultural practices and some timber extraction. They, too, have relatively large expanses of land under traditional management systems, which also means they have been preserving large expanses of tropical rain forest. The mestizos at the present time must be divided into two categories: those with a long history of living on the Atlantic coast—they were either born there or moved there when young—and recent migrants. The traditional coast mestizos clearly regard themselves as costeños while the new migrants (frequently called "spaniards" by locals) bring with them the baggage of Pacific coast Nicaraguan culture.

Presently national politics are driving much of what is happening on the Atlantic coast, and to the rain forests. Current national government policies emphasize privatization and "free markets," undercutting much of what was gained by the peasant class during Sandinista times. Much land is again being purchased for speculative purposes, which makes its price inflate rapidly. Peasant owners in the Pacific regions of Nicaragua can hardly be expected to withstand the pressure of selling their land when it appears they are getting much more than they thought the land was worth. Furthermore, with the idea that there is free or almost free land to be had on the Atlantic coast, the temptation to sell out and move to the Atlantic coast is frequently irresistible.

Fortunately, much of the "free" land that is the traditional land of Creoles and Miskitos is still forested. However, it does not take a very complex analysis to predict that there will be conflicts among those who regard the same piece of land as either "free" or "traditionally ours." For example, on a trip up the Kurinwas River in 1995, our Miskito guide was showing us the forested banks of the river that constituted the traditional lands of the Miskito village of Tasba Pauni. Suddenly, we came upon a clearing, and as we approached the shore, we could hear the telltale raps of axes and machetes clearing forest a little further on. Our guide insisted on confronting whoever was clearing forest, so we moved on with him to eventually find members of a small family, two men and a woman, in the process of clearing the land. The conversation went something like this:

> Miskito guide: What are you doing here? Don't you realize this land belongs to the Miskito people of Tasba Pauni?
>
> Farmer: We didn't know. It is just forest, no one is using it.
>
> Miskito guide: Of course it is forest. We use it as a forest. The Miskito people have been hunting in and extracting timber from these lands for hundreds of years. You have no business cutting it down.
>
> Farmer: But we have nowhere to go. We have no land, there are no jobs to be found. We had to sell our land in Chontales [on the Pacific side], and we were told there was plenty of land to be had here on the coast. How are we supposed to know this is your land? It is just forest.

Miskito guide: But it is our forest and we want it to stay forest. You have to stop your clearing right now.

Farmer (looking nervous): But where are we supposed to go? We have to eat. Our children will starve if we don't plant something.

Miskito guide: Maybe you can find some land in Pueblo Nuevo [a small settlement nearby], but you cannot continue here. This is our land, and we will defend it. We have guns, and we are ready to use them to protect our land.

Farmer: What will become of us?

This is all paraphrased, but the sentiments were much as these quotes suggest. What is difficult to put to paper is the feeling one had in this situation. The farmers were sad indeed, very poor people just trying to eke out a living. The Miskito guide, also a poor man, was defending traditional resource use policies that had been part of his community ever since anyone could remember. If these mestizo farmers were allowed to clear the forest, they wouldn't be able to produce there for very long anyway, and the riches of the forest would be lost forever. But as the mestizo farmer said, "What will become of us?" How would he feed his children? And would the Miskito come back with armed guards? It does not take much reflection to see the incredible potential for violence here. The farmers, if they are smart, will start to organize together to clear larger tracts of land simultaneously. Faced with the prospect of armed Miskito resistance, will they seek arms themselves? Large groups of armed and desperate people with conflicting interests does not make for a stable political climate. Indeed, violence flared early in 2004 when armed Miskito groups traveled through areas recently colonized by new migrants, threatening them with attacks if they did not leave their new homesteads,[18] apparently attempting to send a message to future colonists. Whether the new homesteaders will themselves arm and fight the armed Miskitos is yet to be seen.

The fate of the rain forests of Nicaragua is ultimately tied up with the contentious issue of autonomy. The Atlantic coast people have long held a desire for autonomy, but the political realities of the coast have usually been otherwise. However, during the 1980s when the Sandinistas were in power, a long series of discussions and debates finally resulted in a constitutional amendment that granted the Atlantic coast limited autonomy from the national government. Especially important to the coast people was the right to make their own decisions about their local natural resources, which include all the rain forests of Nicaragua. When the Sandinistas lost the elections in 1990, the coast people effectively lost their influence in high places. Despite the formal recognition of autonomy, the elected regional councils have been extremely weak since their establishment in 1990. Now with the growing hegemony of the PLC and the transformation of the Sandinista party, prospects for the Atlantic coast people's autonomy seem ever more fading.

Yet the question of autonomy is still at the core of politics on the coast. The formalization of a state of autonomy in 1990 was viewed by most costeños as the beginning of a new era; the dignity of local people would be recognized and their power acknowledged through the ballot box. Most importantly, their natural resources, including the rain forests, would be under their control. Their subsequent frustration at the empty promise of real automony has led to new political sentiments.

Autonomy for the Atlantic coast is anathema to the political party currently in power nationally, the PLC. Seeing the riches of the coast's resources, the national administration is not likely to relinquish control over the region. Logging concessions are sold in defiance of the will of the regional council, fishing companies routinely overfish the marine banks, and new migrants from the Pacific side—driven or lured off their own lands by government policy—cut rain forests to make their small farms.

The neoliberal model imposed on the nation in the early 1990s and enthusiastically adopted by the current PLC administration—with the apparent complicity of the Sandinistas—has been devastating in its consequences. Privatizing lands in the populous western half of the country has made it difficult, if not impossible, for small-scale farmers to make ends meet. They are losing their land in record numbers. Many migrate to Managua to contribute to the swelling shantytowns there, but many also see the Atlantic coast as a new opportunity. In this way neoliberal policies are pushing newly landless peasants eastward towards the agricultural frontier. Much—probably most—of the land available on the coast is traditional communal forests held by indigenous and Creole populations. Since the new migrants are culturally mestizos, it is not surprising that deforestation along the new agricultural frontier is attributed to them. Furthermore, since the mestizos are the largest and fastest growing of the ethnic groups in the area, the agricultural frontier has also become a cultural frontier, in which traditional notions of what it means to be costeño is being altered by the new migrants. While most people distinguish between newly arriving migrants and those with a long history on the coast, this distinction is increasingly blurred in recent times. It is easy to see how under the circumstances indigenous and Creole people could come to see the mestizos as their enemy.

It is convenient for the PLC to play up this new ethnic division. Their base of support in the Atlantic region is in the mestizo population, especially among the new migrants. As the flood of migrants increases, support for the PLC grows, especially if mestizos see their position on the coast as threatened by the other ethnic groups. Interviews with mestizos in the late 1990s and early 2000s revealed a new fear on their part. Repeatedly we heard people claim that if autonomy ever became a reality, the Creoles and Miskitos would throw all the mestizos out of the coast region; that the Creoles and Miskitos don't want autonomy, but rather independence. Interviews with Miskitos and Creoles indicated a parallel antipathy towards the mestizos, who are seen as bolstering the hegemony of the central government by voting for the PLC, and as destroying the forests and fishing grounds with their unsustainable agricultural and fishing practices. Furthermore, Creoles and Miskitos seem to be reinventing old rivalries as their political alliances become identified with ethnicity rather than autonomy. In short, ethnic rivalries among the three main ethnic groups seem to be taking the place of the previous politics of autonomy. "My enemy is in a different ethnic group" seems to be replacing the notion that "my enemy is in Managua." All this benefits the PLC, injures the cause of autonomy, and undermines the political will of the groups that have traditionally protected Nicaragua's rain forests.

Atlantic coast intellectuals recognize the growing ethnic tensions and cynicism with traditional politics among costeños, and are warning of the early signs of ethnic conflicts. Researchers at the Center for Research and Documentation of the Atlantic Coast (CIDCA) are emphasizing the importance of coast people identifying as costeños rather than as members of ethnic groups, and of the need for the national government to recognize growing ethnic tensions as an outgrowth of its anti-autonomy position. Workers at the Center for Human, Civil, and Autonomous Rights (CEDEHCA) continue pushing for a coast identity independent of ethnic identity, with autonomy as a central rallying cry. Responsible leaders are trying to reinvigorate the autonomy process as the only way to unify coast people.

What the future will hold is hard to predict. Ethnic warfare among Atlantic coast people is certainly not inevitable, yet it is not out of the question at this point, as it was in the 1980s. But certainly, the future of Nicaragua's rain forests is at least partly tied to the questionable future of the coast's autonomy movement.

Costa Rica: Are Bananas Losing Their Charm?

The situation in Costa Rica has become considerably more complicated than it was in the 1980s and early 1990s. Here we have a good news/bad news situation. Deforestation of tropical forest has declined somewhat according to all estimates, and there are some indications that reforestation on some abandoned agricultural land is becoming important. Furthermore, in some cases, including that of the community of El Progreso (which we discussed earlier), the lives of the people have considerably improved and it is highly unlikely that they will engage in the forest-felling activities that their counterparts did only ten years ago. That is the good news. The bad news is that the near future may bring a conjunction of forces that could change those two tendencies.

There seem to have been two forces involved in Costa Rica's lowered deforestation rates (and its higher reforestation rates). First, the availability of forests not included under protected status is very low, since former high rates of deforestation have resulted in almost complete deforestation outside of the protected areas. Since agricultural land seems to be in the process of being abandoned, much of the pressure on forested lands is lessened. While there is still concern that landless peasants will begin moving into the protected areas, that force seems minimal at the present time. Second, and probably more important, Costa Rica's industrialization program has taken off, and there has been a large rural-to-urban migration in response. Intel, for example, has taken a major role in the development of the technology sector in Costa Rica, which now dominates the country's exports. This rapid development of the industrial sector has created nonrural jobs that attract former peasants from the countryside into the cities. Traditional pressure on the forests is minimized by this migration, as we described earlier in the case of Puerto Rico.

However, this demographic shift created a different problem, at least in the short term. Bananas and coffee are still major sources of income for the country, and both are quite labor-intensive. Not only are they labor-intensive, they require very cheap labor—not the sort of salary scale that industrial workers normally command. Industrialization thus could have created a crisis in Costa Rica's banana and coffee industries. However, the pattern that was beginning in 1994 has accelerated dramatically, with Nicaraguan migrants streaming into the country at rates that would make any banana company official ecstatic. The vast majority of banana and coffee harvesting is now done by Nicaraguans, who also increasingly fill most of the lower-status jobs that Costa Ricans no longer want, such as maid and janitorial jobs. So while the industrial program has elevated the salaries of the average Costa Rican worker, the massive migration from Nicaragua has averted a crisis in the traditional agricultural economic base.

Several additional factors are currently shaping the interplay between poor rural people and the land. One is the abandonment of the enclave system by the banana companies. In the past, when workers were let go, they not only lost their jobs, they lost the roof over their heads as well. For many banana workers this is no longer the case. For example, in the homesteading community of El Progreso, described in detail earlier in this chapter, almost all men and many women are employees of one of the local packing plants. Buses owned by Dole and Chiquita stop by the community to pick up the workers every morning and bring them back in the afternoon. When the packing plant is forced to lay workers off, rather than being forced to seek a piece of land on which to homestead, the workers have already gone through the process of homesteading, and when they lose their jobs they at least still have a roof over their head. For most of them the amount of land they have is probably not sufficient to maintain their families with subsistence production. But it is likely that the inexorable pressure to find a piece of land on which to homestead will be far less evident for these people than it was for earlier enclave workers who, as soon as they were fired, were also kicked out of their houses.

Another factor potentially shaping land use in Costa Rica is the ability of some of the organized workers to take advantage of the country's remarkably progressive social security network. Again, taking El Progreso as an example (we think it is not atypical): just four years after being established the community organized to petition the government for more than the title to their land. (While most members of El Progreso are still waiting for formal titles to their land, there is no question that their right to possession of the land is secure. This is a clear consequence of their initial collective effort in invading the land.) The community organized to build a school for its children and then petitioned the government to send teachers, which it was required by law to do. Sporting activities and community fiestas, which sometimes serve to raise money for community projects, are organized on a communal basis. Visiting the community in 1999, we almost did not recognize what had in 1994 been a ramshackle collection of huts mostly made out of scrap wood. Many of the houses are made of cement blocks; the main road is not yet paved, but it is graded much better than the two-track rut it had been; and small stores dot the landscape. El Progreso now looks like a prosperous community compared to the way it looked in 1994. And it must be noted that this prosperity is built on the capital provided by the banana companies, of which El Progreso residents are seemingly quite supportive.

It is worth reflecting on what caused this remarkable transformation of the El Progreso community. On one hand, there are political conditions in Costa Rica that allow such progress. On the other hand, people are organized to take advantage of those conditions. Both factors are important. First, Costa Rica is not El Salvador or Honduras. Since its liberal revolution in 1948, Costa Rica has had no army, and basic human rights of the lower classes have been respected (at least compared to other countries in Central America). Legal protection of homesteaders extends to all citizens, and tradition accords those protections, to some extent, even to foreigners (i.e., Nicaraguans). When poor people are accorded such rights, they naturally tend to bond together in organizations that can take advantage of those rights. Several civic organizations have emerged in El Progreso, and they function to pressure the government to intervene on behalf of the people. Gravel roads have replaced mud ruts, electricity is available to all residents at a nominal charge, and state-paid teachers instruct the community's children even though none of the residents yet has formal legal title to the land! As Wright and Wolford[19] note of the Landless Workers' Movement (MST) in Brazil, rural organizers are able to help local people force their government to obey the law.

At the same time, working conditions for the banana workers remain more or less the same as they were earlier, with some improvements. Costa Rica has progressive labor and environmental laws and the banana companies have been forced to clean up much of their act regarding housing, pollution, and pesticide exposure. The companies' tendency to hire and fire workers to save on payments to social security seems to have been replaced by a desire to maintain trained workers: the money saved on having to pay social security apparently is small compared to the cost of continuously training new workers. Yet there is still an effective ban on union organizing, which is ultimately the only real power workers have to determine their own conditions of work.

The basic pattern of bust and boom for the banana business has continued. After the massive expansion of the early 1990s, which was interrupted by only one major downturn in prices (in 1992 to 1993), the companies are facing what looks like a major crisis in the near future. According to one Chiquita representative, an individual plantation must be able to sell a box of bananas for $5.80 to break even. In August 1999 they were receiving $4.80 per box, and by 2003 the price had dropped to $3.50. Overproduction, the major problem in all of modern agriculture, seems to have reared its ugly head again, and too many bananas are flooding a market that was *not* stimulated by expansion into eastern Europe, as had been expected in 1989 and 1990. According to this representative, something is going to have to give soon. A recent mammoth expansion of production in Ecuador is currently coming onto the market, and because of lax labor and environmental laws there, competitor companies are producing bananas at $2.00 per box. If these figures are accurate, it is difficult to see how the three major companies (Chiquita, Dole, and Del Monte) can maintain their current high levels of production in Costa Rica. What then will happen to the truly enormous number of workers who have been drawn to work in the plantations?

A hopeful answer to this question is that they will be able to participate in Costa Rica's industrial expansion. But that may not be so easy. Fueled by both the banana expansion and the horrible conditions in Nicaragua, extensive immigration of Nicaraguan workers has occurred since 1990, and it has accelerated since 1995 (the year the first edition of this book was published). Depending on who you listen to, the current population of Costa Rica ranges from 10 percent to 33 percent Nicaraguan. Imagine if, in the past few years, over 100 million Mexican workers had immigrated to the United States—if one of every three people you met were from a different culture. Antagonism against Nicaraguans in Costa Rica is on the rise; spray painted graffiti demands "Nicas go home."

These people, pushed from their homeland by the dire circumstances there and pulled by the lure of the banana companies in Costa Rica, are now doing all the menial jobs that the newly industrialized Costa Rican workers no longer will do (at least not at the wage levels required for international competitiveness). It is not uncommon, for example, to read in the classified sections of local newspapers "Nicaraguan maid wanted." According to Chiquita officials and local union organizers, between 80 and 90 percent of all banana workers are Nicaraguan at the present time. And the second most important export crop, coffee, is now also almost entirely dependent on Nicaraguan labor.

In a sense, the availability of Nicaraguan workers has saved these two most important export crops for Costa Rica. With the industrialization process proceeding at a rapid pace, rural workers have transformed into industrial workers who demand wages that are excessively large for traditional banana or coffee production to be internationally competitive. As industrial output grows, this trend will continue. Indeed, bananas and coffee would surely have already contracted dramatically if they had had to rely on the newly expensive labor.

What can be expected in the near future, when bananas and coffee do contract? Will the Nicaraguans return to Nicaragua when rural jobs evaporate? Perhaps. But of the approximately twenty families we talked with on Chiquita plantations and the surrounding area in August of 1999, none indicated a willingness to return to Nicaragua other than for short trips to visit with family. Will the Nicaraguans move into the industrial sector? Perhaps not, because they are generally unskilled. But skills can be rapidly acquired, and it is not clear exactly how much education is necessary to work on a computer chip assembly line. If they move into the industrial sector, this will exert downward pressure on industrial wages, undercutting the social contract normally associated with the process of industrialization (i.e., rising wages, which create more demand, thus leading to economic growth). Such a scenario could undercut the industrial growth currently underway. Or will the Nicaraguans become the new flexible peasant class, eking out a subsistence living on recently cleared rain forest land until bananas (and coffee) undergo their next phase of expansion?

The Lessons of the Costa Rica/Nicaragua Comparison

That the rain forests of Central America are being destroyed is a point on which there is no debate. There is much debate over the causes of their destruction, however. Many biologically oriented analysts tend to evoke the simple metaphor of the rain forest as a commons, with overpopulation the driving force in its destruction. This analysis might lead to the (erroneous) conclusion, for example, that the development of effective birth control programs in the Sarapiquí would stop small farmers from seeking farms in the area. The reality is different. Since workers are actively recruited into the area by the banana companies, condoms are unlikely to solve the problem of rain forest destruction. Indeed, as far as the banana producers are concerned, the area remains underpopulated.

The alternative explanation, more complicated and nuanced than one based on overpopulation, has been presented in this book. It takes its framework from several classic ideas and sees rain forest destruction as the consequence of the interaction among several forces operating within a particular world system. Three modes of production (forestry, peasant agriculture, and modern agriculture) interact with two socioeconomic groups (export agriculture bourgeoisie and rural peasantry or proletariat) in the context of national and international political structures and within a complicated ecological matrix. It is impossible to understand deforestation if one remains at only one or another of these levels, since the cause is located within the overall structure.

A comparison between Nicaragua and Costa Rica during the early 1980s is illuminating. Both countries had significant areas of rain forest remaining, but Costa Rica lost rain forest areas at an extremely rapid rate during the 1980s, while the rate of deforestation in Nicaragua was much lower. How can we understand this difference?

Nicaragua was distinct from the other Central American countries during the 1980s. For one thing, the contra war militated against significant lumber operations, hence extensive access roads were not built. The war, while sometimes intense in places along the northern border, was a more low-intensity conflict where the rain forests were concentrated on the Atlantic coast. Small bands of counterrevolutionaries were scattered throughout the Atlantic lowlands, especially in heavily forested areas. Thus the region was not likely to be subjected to intense commercial logging—or even small peasant clearings—during the war.

More importantly, however, the massive agrarian reform program initiated after the 1979 revolution all but eliminated hunger for land among peasants in Nicaragua. In 1978, 36 percent of the country's land was in farms larger than 850 acres. By 1985, that fraction had dwindled to 11 percent. In 1978, there were no production cooperatives. By 1985, 9 percent of the land and 50,000 families were integrated into the cooperative sector. And by 1985, over 127,000 families had received title to their own land.[20] This was an agrarian reform without precedent in the history of Latin America.

Costa Rica's agrarian reform program, organized by the state Institute of Agrarian Development (IDA), was entirely different, bearing the recognizable fingerprint of similar programs in El Salvador, and—years before that—in Vietnam. Landless peasants were located in or near undeveloped areas—only occasionally on the lands of large landowners. Expropriation of large, inefficient enterprises was largely absent from the system, and land titles were given with twenty- or thirty-year mortgages, forcing farmers into cash crop production. Traditional farmers who sought a plot of land acquired a piece of marginal, undeveloped land (if they got anything at all) and, for the first time in their lives, a bank debt. Exceptions to this basic pattern existed (e.g., the El Progreso community as described earlier), but the majority of small-scale producers lived in this sort of politically marginal zone.

Interviews with small farmers in both countries reflected the basic differences in agrarian reform programs. Costa Ricans emphasized the land tenure issue, voicing disquietude over their lack of land title or their inability to pay recently acquired mortgages (many were quite surprised at their new debt), and focused their economic concerns on the attainment of land security. Nicaraguans were the opposite, at least before the 1990 elections. While they had many legitimate gripes, lack of land was not one of them, and rarely did a Nicaraguan small farmer talk about needing a piece of land to call his own. The consequences of each system with regard to pressure on rain forests are obvious.

For example, one could have predicted that the pattern of land-hungry peasants following logging roads into the forest, so common in Costa Rica, would not be a problem in Nicaragua. But the electoral changes in Nicaragua in 1990 kept the country's new model of land tenure from being put to the test, since the new government made the rollback of much of the agrarian reform program part of its electoral platform. As expected, and as detailed earlier in this chapter, the new class of landless peasants created by this rollback is busy following new logging roads and clearing forests in eastern Nicaragua.

One point of comparison outside of Central America is also worth making. The Caribbean nations of Puerto Rico and Cuba are the only tropical countries in the world to have experienced an increase in the area covered by forest in the past few decades,[21] though for different reasons. Because of the development program imposed by the colonial administration in the 1940s, Puerto Rico can hardly be considered an agrarian economy. Because of its peculiar articulation with the United States and in a grotesque parody of real development, Puerto Rico has become an industrial colony of the U.S. Consequently, to speak of a rural peasantry in

Puerto Rico is anachronistic, and any notion of a movement of landless peasants into forested lands is ludicrous. Industrial development, however artificial and imposed undemocratically from afar, has transformed Puerto Rican society so much that the basic class structure that causes deforestation simply no longer exists there, and the normal course of forest succession is taking place.

Cuba's recent development, which emphasizes socialist goals, has been along different lines than Puerto Rico's. The successes and failures of Cuba's line of development are obviously hot topics for debate, but a single feature of Cuban developmental strategy is relevant to the present discussion. Cuban peasants are not land-hungry. Because of the government's basic philosophical commitment to a secure economic environment for all members of society, the Cuban peasantry has either been absorbed into the urban sector or has received land security on state farms, cooperatives, or private farms. Cuba simply has no landless peasants.

Clearly, the probability that either the Cuban socialist solution or the Puerto Rican federal-subsidy solution will occur in Nicaragua, Costa Rica, or any other tropical country is vanishingly small. So what might we realistically expect in Central America? Indeed the prospects do not seem bright. If general world trends continue—and especially if the relationship between the developed and underdeveloped worlds evolves along the lines anticipated by the so-called neoliberal new world order—we can only expect more of the same. The class conflicts that raged in Guatemala, El Salvador, and Nicaragua during the past twenty years are not likely to subside, although their specific form may change. With the threat of U.S. military adventure constantly on the horizon—and with the U.S.'s reach no longer limited by the Soviet Union—we can expect to see continued increase in the power of bourgeois elements in these economies and consequent further erosion of the land base of these countries' peasantry.

Every factor seems to be in place for the continued growth of landlessness and poverty in Central America (and elsewhere in the developing world). If industrialization cannot absorb the expanding landless peasantry, then a replay of past deforestation patterns seems unavoidable. And with the current international debt situation and continuing political conflict in many countries in the region, the prospects for significant industrialization do not seem very bright. The vision of agrarian reform programs like Nicaragua's in the 1980s has all but disappeared in its country of birth, and is hardly a serious proposition in the other countries of the region. Thus, the basic programs that would help stem the tide of deforestation, agrarian reform and industrialization, seem equally far off.

This puts conservationists into a holding pattern dividing along the same lines we introduced in Chapter 1. Mainstream environmentalists have been concerned with accumulating large sums of money to purchase and protect islands of pristine rain forest, with little concern for what happens to either the ecosystems or the human societies located between those islands. As we indicated previously, we do not imagine this strategy has much chance of working in the long run. The landscape will be converted into isolated islands of tropical rain forest, drowning in a sea of pesticide-drenched modern agriculture, with masses of landless peasants looking for some way to support their families. Unless the pace of industrialization increases dramatically (something few observers expect), these peasants are unlikely to be absorbed into the industrial workforce. Instead, they are very likely to begin homesteading in those islands of protected forests.

The alternative strategy focuses on the land *between* the islands of protected forest and acknowledges the interconnections in this complicated system. This sustainable development perspective conceives the ecological side of the dilemma as a landscape problem, with forests, forestry, agroforestry, and agriculture as interrelated land use systems, and seeks to develop those land use systems to maintain conditions of production. This strategy has seen much recent analysis,[22] and is perhaps our best hope.

Past historical contingencies and current economic realities place the Global South in a disadvantaged position for economic development under the world system. Development specialists from all spheres of political and economic influence have attempted to deal with the problem of third world underdevelopment ever since it was recognized, but economic development, irrespective of its sustainability, has not been common in the third world. Whatever restrictions are imposed on the development process only heighten the economically disadvantaged position of the Global South. Indeed, under our current world system, an economic advantage will accrue to those regions and countries that do not impose constraints on their development. Sustainability is just such a constraint—from the short-term perspective.

In the End, a Pessimistic Assessment?

In the first edition of this book, we ended this chapter with a paradox and with a note of pessimism. We now feel there is a suggestion of optimism that can be put forward. We'll begin by repeating the pessimistic conclusion we drew in 1995, and then provide a somewhat more optimistic assessment in the subsequent section.

Stemming the tide of rain forest destruction requires not only development, but development that is sustainable. Obviously, sustainability is not sufficient, but it is a necessary requirement. In the current world order, rational planning is not anticipated at either national or international levels; development will proceed fastest for those countries able to ignore constraints that others either cannot or will not ignore. As we have noted, sustainability is precisely such a constraint. Thus, although sustainability in development is required to save the rain forests, under conditions of the current world order it seems unacceptable as a constraint on short-term development.

Our note of pessimism (if the above paradox is not sufficiently pessimistic) is that the structures to which we lay blame for the destruction of rain forests are still in place—and probably more solidly than they were ten years ago. The same modes of production (forestry, peasant agriculture, modern agriculture) interact with the same socioeconomic groups (export bourgeoisie, peasantry) in the same national and international political arena within the same ecological matrix. Forces in the national and international political arena have constructed an ideology in which the rain forest is considered an externality that fits into and fortifies the overall political structure, thus creating an inexorable dynamic of which deforestation is an inevitable consequence. As long as present political arrangements survive, a solution simply does not seem possible. To the extent that conservationists and the conservation movement are part and parcel of those political arrangements, they remain part of the problem.

In the End, a Note of Optimism?

The world has generally continued on the course that led us to our pessimistic assessment, especially with the bellicose attitude adopted by the United States subsequent to the attacks of September 11, 2001, which we expect to continue or even intensify during the second term of the Bush administration. Nevertheless, as we indicated in Chapter 6, there is another part of the world that seems to be developing differently. As noted by Hardt and Negri in their perceptive book *Multitude*, new global political arrangements are rapidly evolving, spurred on by world travel and advanced telecommunications, including innovations like the Internet. Rather than seeing the United States and whatever coalition it imagines dominating the world stage, these authors focus on the new forms of social organization that seem to be emerging parallel with, and very much contrary to, the dominant nationalistic paradigm that U.S. political culture promotes. Theirs is a sophisticated and complicated analysis that cannot be repeated here, but suffice it to say that the political arrangements that we argue cause rain forest destruction seem to be breaking down, and the important international coalition is not the so-called coalition of the willing crafted to justify the U.S. invasion of Iraq, but the coalition of the multitude, the collection of grassroots organizations that claim that people's basic needs are to be given priority over the needs of the international capitalist system. Operating without a hierarchical leadership, according to Hardt and Negri, gives this movement a long-term strength and a potential for evolving a new mode of human organization that looks more like a network of interconnected nodes operating more or less independently of one another, but with a common overall goal. Indeed, in the first edition of our book we saw a great deal of gloom on the political horizon, with the neoliberal model seemingly penetrating all corners of the globe and dominating all aspects of life, promoting and strengthening precisely those aspects of the world system that we saw creating rain forest destruction. But since that time we have seen the Battle of Seattle and subsequent massive protests over the capitalist globalization scheme; the coming together of labor and environmental groups into strong coalitions; and the rise of sensible approaches to agricultural development and energy use. In short, the new global movement for social and environmental justice seems poised for a historic victory. If so, our remaining rain forests may survive.

9: BIODIVERSITY, AGRICULTURE, AND RAIN FORESTS

DESPITE THE FACT that there exists great potential to find new products in rain forests, possibly even cures for diseases like cancer, it is not really these utilitarian notions that fire the public imagination. Rather, it is the amazing fact that rain forests contain a disproportionate amount of the biodiversity on the planet, as discussed in Chapter 2.

Even disregarding all the potential rain forest products, rain forests' possible use as carbon sinks,[1] and the many other useful things that have been imagined for or from rain forests, observations about rain forest biodiversity cannot help but strike a deep chord. Although rain forests cover only about 7 percent of the surface of the earth, they are thought to contain more than 50 percent of its biodiversity.[2] Anyone would respond with deep emotion if they learned their hometown—inhabited by their mother, father, brothers, sisters, aunts, and uncles—was about to be visited by a natural disaster. Perhaps a similar deep feeling is stirred when we note, as biological creatures, that the majority of our living relatives are in the rain forest, and that rain forests are being rapidly destroyed.

In a sense this is the ultimate concern. We must acknowledge all the potential uses of this biodiversity, but it is important to recognize that a deeper principle is involved. If some force were destroying the world's art museums, all thinking humans would be concerned. Such a force is, right now, destroying treasures that are perhaps even more precious and irreplaceable than the contents of the world's art museums. Our concern is justified.

Biodiversity and Utilitarianism

Still, there is no doubt that a utilitarian focus on biodiversity is to some extent valid. While the ultimate reasons for concern over the loss of biodiversity may not fall within the category of utility, the historical record and common sense both suggest that biodiversity indeed has value as a source of useful things for humanity.

The last five centuries certainly make this clear. Much of the initial economic push that led to European conquests of non-European lands came from the search for products from tropical lands. The European desire for spices drove the indefatigable search for the almost mystical spice islands, and accounts to some extent for the early Dutch experiments in colonialism. The English obsession with tea sweetened with sugar arose from England's domination of colonies in south Asia and America. Recall the ultimate purpose of Captain Bligh's journey—a Pacific voyage for breadfruit trees to feed the slaves in Britain's Caribbean colonies. Bananas, chocolate, coffee, tomatoes, and many other food and drug crops that today we take for granted have their commercial origins with entrepreneurs, pirates, and governments seeking to make use of tropical biodiversity. In fact, tropical rain forests contain many of the wild cousins of domesticated food and industrial plants. In addition, many modern medicines come from plant material extracted from the rain forest.

It would be folly to suggest that this historical pattern has suddenly come to an end. Indeed, as we write, pharmaceutical companies from Merck to Bristol-Myers Squibb are searching the remaining patches of tropical rain forests for products they can market. The centerpiece of Costa Rica's National Institute for Biodiversity is a search for products that can be sold to pharmaceutical companies.

But there is another component to the biodiversity question that is sometimes lost. When bird-watchers first come to a tropical rain forest area, they are frequently quite disappointed. The birds are difficult to see, largely due to the lush vegetation that provides them with cover. Thus one sees the amusing irony of bird lovers who, having visited a tropical country to see the beautiful birds of the rain forest, seek out whatever patch of open, non–rain forest area they can find in order to actually be able to see their birds. Indeed, in any rain forest reserve in the world, if you wish to spot bird-watchers, look in the abandoned agricultural areas.

Furthermore, recall from Chapter 2 that disturbance—sometimes modest, sometimes catastrophic—is a perfectly natural component of the rain forest cycle. So the animals and plants of the rain forest do not require an untouched, cathedral-like green mansion, but rather expect a mosaic of dark understory, light gaps, old landslides, tornado and hurricane tracks, and the like. There is a parallel to this concerning the various forms of human activity in rain forest areas. Some activities are so minor that most organisms in the forest are unaffected, while others are so dramatic that the recovery of the forest is delayed for a very long time. We categorized the types of damage in Chapter 6 according to rates of recovery. But what about that most critical of all the features of the rain forest—its biodiversity?

This problem is usually approached by asking, "What would happen to biodiversity if all the rain forests were cut down?" It seems like a straightforward question, but it is grossly misleading. Indeed, little is known for certain about what happens to biodiversity when logging occurs. The image fixed in most people's mind is one of devastation: thousands of species replaced by a field of tree stumps. But this is not the consequence of timber harvesting per se. The agricultural activities that almost inevitably follow a logging operation are far more devastating than the logging itself. Since the forest is usually resilient to major damaging events, absolute loss of species as a direct result of logging is probably rare. The trees resprout, the birds hide in the resprouting trees, and who knows what happens to the insects and other smaller animals?

In fact we have little evidence that the simple act of cutting trees does anything whatsoever to biodiversity. Of course the nature of the logging operation itself is likely to have an effect, and certainly its extensiveness is also likely to have an effect. But on the evidence of the few studies available,[3] we really must withhold judgment. It may well turn out that, despite the devastated physical appearance of a hectare of logged forest, little biodiversity loss actually occurs due to the *direct* effects of a logging operation.

This surprising observation leads to a point made earlier. It is what happens *after* the rain forest is cut down that matters. If you cut it down and put a city in its place, clearly you have changed the level of biodiversity for a long time; perhaps forever. But if you cut down a few trees and go away, the trees resprout, ecological succession proceeds in the light gap produced by the cut, and there is little evidence that biodiversity is reduced at all. There is obviously a gradient between the cutting of a few trees and the construction of a parking lot. The rate of recuperation of the forest will be extremely rapid at one end and extremely slow at the other. But more importantly, the loss in biodiversity will be small at one end and large at the other. We must now ask a serious question: exactly how much biodiversity will be lost through various modifications of the rain forest environment? Surprisingly, despite numerous emotional pronouncements, little is known to answer this question.[4]

Biodiversity and Agriculture

Recent focus on the preservation of biodiversity is based on the suggestion that more species became extinct in the latter part of the twentieth century than at any other time in history. While efforts to curb these losses have intensified in recent years, emphasis has been on the preservation of a few charismatic and conspicuous organisms, or on creating pristine environments, mostly within national parks and reserves. In fact such organisms are a very small fraction of the threatened biodiversity, and such habitats represent only a small percentage of total land area. While obviously on a unit area basis such habitats are the most biologically diverse in the world, it is also true that managed ecosystems are far more extensive in area. Since the combination of managed ecosystems and human settlements cover approximately 95 percent of the earth's terrestrial surface, it might be argued that it is as important to examine patterns of biological diversity in managed ecosystems as it is in highly diverse unmanaged ones like tropical rain forests.[5]

For example, with seventy-four conservation units (including national parks, biological reserves, and national monuments) covering 1,154,945 hectares,[6] Costa Rica has one of the world's highest proportions of land under protected status. Yet even in this remarkable case, almost four million hectares (73 percent) of the country's total land are covered in agroecosystems, managed forests, and human settlements. Furthermore, while a fraction of that area is in some form of traditional agriculture, the rest has been, or is being, transformed into high-input monocultural systems. Recently we have come to understand that indirect biodiversity losses through agricultural transformation in this system may be very large.[7]

Regarding biodiversity in agroecosystems, the literature is clear on one major point: that there are two distinct components of biodiversity. The first component is the biodiversity associated with the crops and livestock purposefully included in the agroecosystem by the farmer. A traditional home garden is more diverse than a modern wheat field because many crops are planted in the former and only one in the latter. This is "planned biodiversity." While planned biodiversity is perhaps the most visually obvious component, and has received the greatest attention, "associated biodiversity" is at least as important. This component includes all the soil flora and fauna, the herbivorous, carnivorous, and fungus-feeding insects, the birds and mammals, the associated plants (some of which are weeds), and more.

Most discussion of biodiversity in agroecosystems has centered on the function of the planned biodiversity. For example, it is widely believed that a diverse assemblage of crops in a multiple cropping system reduces market risk. Trees in coffee plantations are considered necessary to provide partial shade for coffee plants. Rotating soybeans with corn provides nitrogen to the soil. Many other examples could be cited.

Associated biodiversity is less well understood. When we compare a Javanese home garden with a pesticide-drenched rice paddy, we find a rich diversity of soil micro- and macro-flora and fauna in the former, but not the latter. This is the associated biodiversity. While the function of this component within the ecosystem is debatable, it is affected indirectly through the planned biodiversity of the home garden.

We illustrate this relationship in Figure 9.1. Planned biodiversity has a direct function within the ecosystem, as illustrated by the bold arrow connecting the planned biodiversity box with the ecosystem function box. Associated biodiversity also has an ecosystem function. But associated biodiversity derives from planned biodiversity; thus, planned biodiversity also

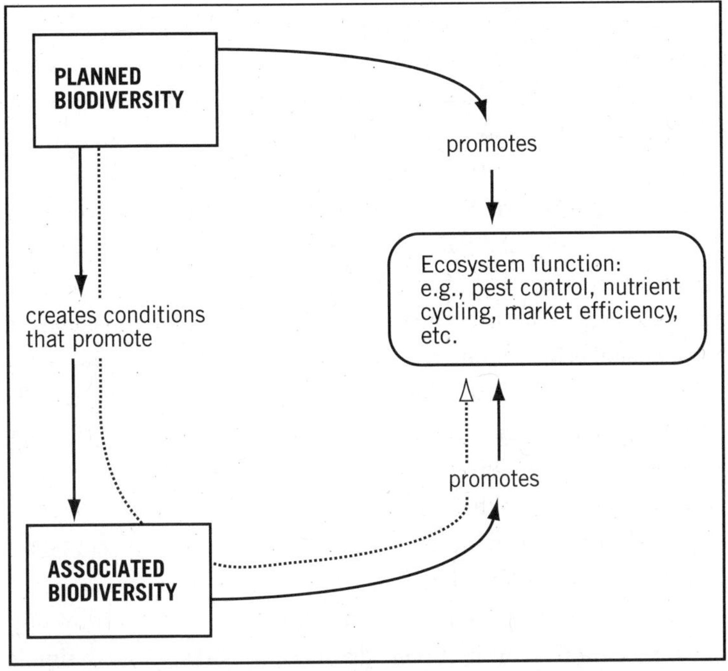

Figure 9.1. The relationship between planned biodiversity (that which the farmer decides to include in the agroecosystem) and associated biodiversity (that which moves into the agroecosystem after it has been set up by the farmer), and how the two promote ecosystem function.

has an indirect function (illustrated by the dotted arrow in the figure) that is realized through its facilitation of the associated biodiversity. For example, the trees in a home garden create shade, which makes it possible to grow some sun-intolerant crops. So the direct function of this second species (the trees) is to create shade. Yet along with the trees might come small wasps that seek out the nectar in the tree's flowers. These wasps may in turn be the natural predators of pests that normally attack the crops. The wasps are part of the associated biodiversity. The trees, then, create shade (direct function) and attract wasps (indirect function).

This simple example notwithstanding, the actual rules that dictate how biodiversity is translated into function are different in managed and unmanaged ecosystems. Farmers not only plan the biodiversity but engage in management practices that may very well alter the manner in which that biodiversity is translated into function, either direct or indirect.

How associated biodiversity is determined by planned biodiversity is a legitimate and interesting question in its own right, independent of the ultimate function of the associated biodiversity. Yet a different set of questions arises from consideration of the continuing changes in human agricultural activities. As farmers change their management practices, they may or may not change the planned biodiversity. Clearly, as the old native multiple cropping system of Mexico is replaced by monocultures of corn, we expect changes in the associated biodiversity also. But even a change in corn monocultures in Central America from a high labor/low capital input system to a system involving high capital outlays also has a potentially important effect on associated biodiversity. The use of pesticides alone may result in an enormous biodiversity loss. Thus, changes in husbandry practices may force changes in the way planned biodiversity functions, the way it translates into associated biodiversity, and in the way the overall biodiversity affects ecosystem function.

Given these observations, we can summarize the relationship between agricultural intensification and associated biodiversity as a simple graph (Figure 9.2). While there is controversy over what actually constitutes intensification, there is general agreement about the extremes. At one end, for example, are the traditional Javanese home gardens, while at the other are the modern rice plantations—paddies prepared with machines, direct seeded with automatic seeders, sprayed with pesticides from airplanes, and mechanically harvested. At one end we find the small patches of corn and beans that Native Americans used to plant in forest openings in Michigan, while at the other end we find the modern wheat fields of Nebraska or Ukraine. It is not difficult to recognize the extreme cases: a patch of corn and beans planted in a natural light gap at one end and a banana plantation at the other. While most agree that biodiversity decreases as intensification increases, the exact form of that pattern elicits significant discussion and debate. In Figure 9.2 we have illustrated what we believe to be the four qualitatively distinct possibilities.

Most ecologists concerned with biodiversity in undisturbed ecosystems have tacitly assumed that the pattern exhibited by curve I is the most likely case: a dramatic loss in biodiversity as soon as any human use and

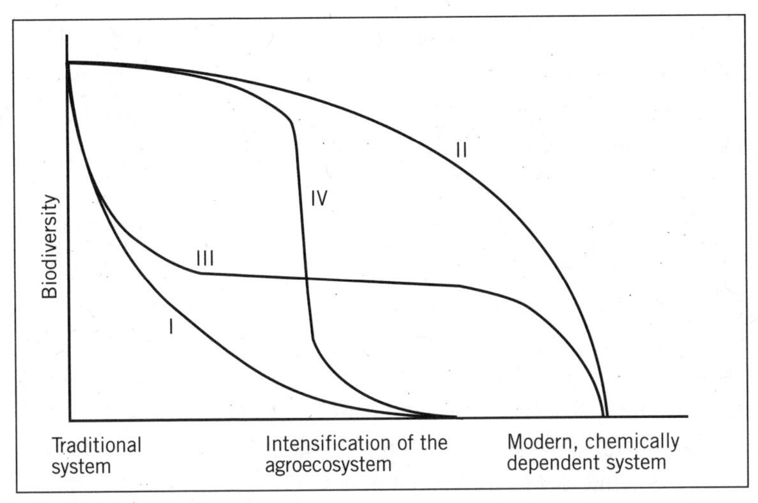

Figure 9.2. How biodiversity changes as a function of the intensification of the agroecosystem, with the four distinct patterns as discussed in the text.

management is brought to bear on the ecosystem. The other extreme, curve II, would perhaps seem unlikely to most ecologists, yet we know of no data or theoretical argument that speaks to the issue. Indeed, the small amount of data available[8] might be interpreted as support for a type II curve. However, we suspect that in most cases something between these two states will be the truth; either curve III or curve IV.

Curve III is a softer version of the ecologist's expectation and simply indicates that, after an initial dramatic loss in biodiversity, the further loss as management intensifies is relatively slight until the extreme of the truly modern systems. Curve IV is perhaps the most interesting case if one is concerned with biodiversity preservation per se. This curve suggests that initial stages of management do little to overall biodiversity, the loss in biodiversity remaining gradual until some critical stage of management intensity. At that critical stage biodiversity declines very rapidly. If biodiversity per se is the concern, and curve IV is an accurate picture, planning activities ought to be focused on maintaining management intensities below that critical point, rather than on aiming at a zero-management strategy. This would mean focusing on sustainably managed agroecosystems rather than pristine unmanaged ones.

Concerns about loss of biodiversity have usually centered on the transformation from natural forest to agriculture, while the transformation from low to high intensity forms of management, which today involves an enormous amount of land, has been practically ignored. It is clear that the greatest biodiversity per unit area exists in tropical forests, and since these forests are being destroyed at such a rapid rate, the bulk of the world's efforts at cataloging and conserving biodiversity are justifiably aimed at these disappearing ecosystems. Yet it seems equally obvious that if, as is sometimes suggested,[9] there exists substantial biodiversity in traditional agroecosystems, then transforming them into modern, intensively managed systems also represents a potentially significant loss of biodiversity. Surprisingly, we know as little about biodiversity in traditional tropical agroecosystems as we know about biodiversity in the tropical rain forest, and have little idea which curve in Figure 9.2 may actually be correct. With a formidable amount of land in the tropics currently undergoing transformation from traditional to modern agroecosystems,[10] it could well turn out that substantial biodiversity is being lost through this change. Clearly this question warrants further study.

Certain kinds of biodiversity attract the greatest attention. Little sparks the public imagination more than elephants or other "charismatic megafauna." Ridicule was heaped on conservationists in the United States when they voiced concern about the extinction of the snail darter, but no one takes lightly the potential extinction of the African elephant. We agree with those who assert that focus on charismatic megafauna helps attract the attention of a public as yet unaware of the problems associated with biodiversity loss in general. There are, however, also problems with this focus. Little in the popular conservation literature suggests that initial concern for charismatic megafauna ultimately translates into any other concerns—indeed the evidence might actually support the opposite outcome.[11] If all available political energy becomes focused on such creatures, little energy remains for concern about the rest of the world's biodiversity.

Small organisms, such as insects, are frequently very specialized and thus are highly susceptible to extinction when vegetation and habitats are modified. Small organisms also represent the vast majority of the world's biodiversity. For example, there are almost one million described species of arthropods,[12] and it is estimated that there are ultimately anywhere from five million to thirty million species of insects alone. In addition, arthropods are potentially important pests, as well as the natural enemies of pests in agriculture. Other groups of organisms are likewise enormously diverse: plants, roundworms, fungi, algae, bacteria, and more. The smallest and least noticeable of organisms constitute almost all of the biodiversity on earth. Nevertheless, most popular concern with biodiversity loss is aimed at the charismatic creature—the elephants, tigers, rhinoceroses, and the like. For the very same reasons that we should focus on agroecosystems when concerned with biodiversity loss, our "organism" focus should aim at the little ones.

The Larger Landscape

In Figure 9.3 we show an aerial view of the landscape as it currently exists near Bluefields, Nicaragua, in an area that is usually considered simply "deforested." What is clear from this figure is that "deforestation" has not been complete. Rather, patches of forest remain within a matrix of agriculture. It is important that we acknowledge that much, if not most, of the current tropical world is fragmented, with patches of rain forest in a sea of agriculture, like the photograph in Figure 9.3. Biodiversity preserves constitute a small sampling of large fragments, but they are fragments nonetheless. The largest biodiversity preserve in the world is Tanzania's Selous National Park (21,000 square kilometers), and even it appears as a tiny fragment on the globe as a whole. National parks and biological preserves may justifiably claim success in biodiversity preservation the world over, yet there remains an extremely large amount of biodiversity outside of preserve areas, scattered in fragmented landscapes from Siberia to Congo. It is thus critical to view conservation at a landscape level, in which fragments of native habitat almost always exist, but are mostly embedded within a matrix of agriculture and/or settlement. There is no reason to view either the fragments or their surrounding matrices as systems that can or should be treated independently by conservationists or developers. Rather we advocate an approach that focuses on the matrix and, following current ecological understanding, the interactions it has with remaining fragments of "natural" habitats.

Figure 9.3. Patchwork of forest fragments in a sea of agriculture, mainly pastures, in eastern Nicaragua, near Bluefields.

When we talked about some of the basic ecological forces involved in rain forests in Chapter 2, we noted the importance of spatial distribution. Recall the so-called enemies hypothesis, in which clumps of individual trees were subjected to attack by diseases or herbivores, thus opening up patches that could be occupied by other species. The key idea is that a clump of individuals would all die, more or less at the same time. In a fragmented landscape, each small fragment is sort of like one of those clumps, and we expect that probably all the individuals in that clump (in the fragment) will occasionally die. Indeed, recent studies have demonstrated that when habitats are fragmented, the likelihood of some species disappearing from the fragment is substantially larger than we suspected.[13] Also recall from Chapter 2 that the dispersal of seeds or migration of animals was an essential feature of the overall dynamics that insured that locally freed patches could be occupied by other species. If we combine this piece of solid ecological knowledge with the fact that most tropical rain forest areas are really small fragments of forest in a matrix of agriculture, we come to the conclusion that migration or dispersal from fragment to fragment is a potentially vital component of biodiversity preservation. It may be that the sorts of fragments of forest that still exist will only be effective harbors of biodiversity if the biological organisms in question are able to get from fragment to fragment over the long term. And this means that the quality of the agricultural matrix—its ability to support associated biodiversity, at least temporarily—becomes a key element in biodiversity preservation.[14]

If the quality of the agricultural matrix matters, we then have to ask the question of how quality is determined, in the sense of insuring a migration pathway from fragment to fragment. This question is obviously related to the question of associated biodiversity as presented in the previous section of this chapter. But it is also subtly different. In this case, we are not necessarily concerned with the ability of the agricultural matrix to sustain biodiversity in and of itself, over the long term. We are concerned with the ability of organisms to get through that matrix so they can migrate from one fragment and colonize another. Sometimes the nature of the matrix from this point of view is quite obvious. For example, deer are not only able to migrate through agricultural areas, they absolutely love those areas, since they eat the produce. However, if we focus on the small things, as we argued earlier ought to be the case, it is difficult to imagine small forest beetles crawling over the surface of the ground in a large soybean field that is drenched with insecticides. We propose, as a rule of thumb, that the agricultural technologies normally championed by the agroecological movement ("organic" or "alternative" or "ecological," etc.) are precisely the sorts of technologies that will allow for the migration of organisms from one patch to another. In addition to the benefits of agroecological techniques from the point of view of food and fiber production, they seem to be also good for the maintenance of biodiversity in large landscapes.

An Example: The Coffee Agroecosystem of Central America

The significance of the above observations can be observed in almost any managed ecosystem in the world. An example in which the problem has been recently evaluated is the coffee agroecosystem of Central America. It is a tropical habitat of the kind found in many of the midelevation tropical areas of the world. In its traditional form it has a forestlike structure and harbors a range of insect communities. For the past decade, it has been

Figure 9.4. The traditional coffee agroecosystem in Central America, aerial view (top) and view from within (bottom). Photos from the Central Valley of Costa Rica.

undergoing a transformation from forestlike to monoculture, and displays clear patterns of biodiversity loss in vegetation (i.e., the planned biodiversity), implying similar loss patterns in the insect communities (i.e., associated biodiversity).[15] In a sense, this habitat represents a natural

Figure 9.5. The modern, chemically intensive coffee agroecosystem in Central America. Entire landscapes are covered with monocultures of coffee (top and middle photos). Bottom photo shows worker, with very little protective gear, receiving pesticide in his backpack sprayer. Photos from the Central Valley of Costa Rica.

experiment in biodiversity dynamics. Replicated experiments have effectively been set up, with some coffee plantations retaining the presumed biodiversity-preserving traditional form and others in some stage of presumed degradation as a result of the modernization process.

As with other major ecosystem transformations in tropical latitudes, the transformation of the coffee agroecosystem involves spectacular landscape changes. The traditional system follows a common pattern of agroforestry, with a variety of shade tree species frequently interspersed with fruit trees, sometimes with relatively dense plantings of bananas in a forest "canopy" above the coffee bushes. The coffee itself is managed at the level of individual coffee plants. Pruning creates small light gaps in which cassava, yams, or other annual crops are planted. When a whole group of coffee bushes are to be renovated (removed and replanted with new bushes), a larger light gap is created and may receive a planting of corn, beans, or other light-demanding crops. Thus, traditional coffee farms share many structural attributes normally associated with forests (Figure 9.4).

The "modern" monocultural system that is being promoted all over the world could not be more different. All the shade trees are eliminated; the traditional coffee varieties are replaced by new sun-tolerant and shorter varieties that are genetically homogeneous; the plants are pruned either by row or by plot, and are heavily dependent on agrochemicals, especially herbicides and fertilizers (Figure 9.5). These two systems represent the two extremes in a continuum of management systems with varying degrees of complexity.

Although such transformation is qualitatively spectacular at the landscape level, little quantitative data have been collected to estimate biodiversity loss. The entomological laboratory of the National Autonomous University of Costa Rica[16] has been doing studies of the arboreal arthropod fauna and has encountered some fascinating results. Arthropods were collected from shade trees, coffee bushes, and the ground. Samples were taken in (1) a traditional coffee farm characterized by a high diversity of shade trees; (2) a more intensively managed plantation; and (3) a "modern" monocultural plantation. These sampled farms represent three points along the imagined intensification gradient of Figure 9.2. As expected, the biodiversity of the arthropod groups declined as the intensification

Table 9.1. Insect species found in the three coffee systems.

Position along Intensification Gradient (see Figure 9.2)	**Traditional**	**Intensively Managed**	**Modern Agrochemical System**
Beetles in shade trees	128	50	0*
Ants in shade trees	30	5	0*
Wasps in shade trees	103	46	0*
Ants on ground	25	14	8
Beetles on coffee bushes	39	29	29
Ants on coffee bushes	14	9	8
Wasps on coffee bushes	34	31	30

* There are no shade trees in the modern system.

gradient increased, as shown in Table 9.1. But more importantly, the diversity of arthropods in the traditionally managed system was impressively high. Indeed, it was almost within the same order of magnitude as that reported for beetles and ants living in the canopies of rain forest trees.17

Other studies have corroborated this basic tendency: the associated biodiversity follows the same pattern as the planned biodiversity. As the system is transformed from its traditional state—three or four species of shade trees, some twenty to thirty species of fruit and lumber trees, plantains, bananas, and understory crops and light gap annual crops—to a monoculture of coffee bushes, the associated number of beetles, ants, wasps, and spiders decreases. In this example it appears that either the type III or type II curves of Figure 9.2 are the true patterns. Yet even in this system—probably the best-studied agroecosystem thus far—from the point of view of biodiversity changes, we are not certain what the true quantitative effects are.

There is, however, evidence that more traditional sorts of coffee production represent a higher quality matrix from the point of view of connecting fragments of rain forest. We recall a conversation we had with a large coffee producer in southern Mexico. We were talking about the Highland Guan, a species know to be restricted to forest habitats. We asked him if it was ever possible to spot one of these in the local shaded coffee plantations, and he responded, "almost never." What struck us was the word "almost" rather than the word "never." That word, "almost," signified that this bird, so dependent on natural tropical rain forest, could nevertheless travel through a shaded coffee plantation. Probably it could never survive there as a viable population, but it could use such a plantation as a migration pathway from one fragment to another. While technical research on this question is extremely rare, there is some evidence to suggest that, at least for small creatures like ants, the more shaded plantations offer a higher-quality matrix, in that migration through them seems more likely.[18]

Models for the Future

The above implies an alternative model for the preservation of biological diversity in rain forest areas. The classical model has been to set up preserves by purchasing the land on which the rain forests sit, and to declare such land inviolable for other human uses. The preservation of biodiversity is viewed as the antithesis of other types of human activities such as forestry and agriculture. The classical model thus strikes a Faustian bargain in which the inviolable lands are to be religiously respected while all other lands can be ignored—let us have our 100 hectare wilderness, and we will remain silent when the other 100,000 hectares of forest are destroyed.

Our alternative model takes into account what we already know about biodiversity not only in tropical rain forest areas, but also in managed systems—agroecosystems and forestry systems. There are two principles upon which the alternative model is based. The first principle is that significant amounts of biodiversity occur in nonpristine rain forest areas such as logged forests and agricultural areas. However, a corollary to this principle, as illustrated by the example of the coffee agroecosystem, is that various types of logging, silviculture, and agriculture imply varying degrees of biodiversity. The second principle is that biodiversity preservation should be part of the planning process in agricultural and logging areas. If, for example, we had known ahead of time the devastating consequences of DDT with regard to biodiversity, its introduction as part of the agricultural system would not have been permitted in the first place. Similarly, the introduction of a new coffee technology should be judged not solely on the basis of the pounds of coffee beans it produces per hectare this year, but

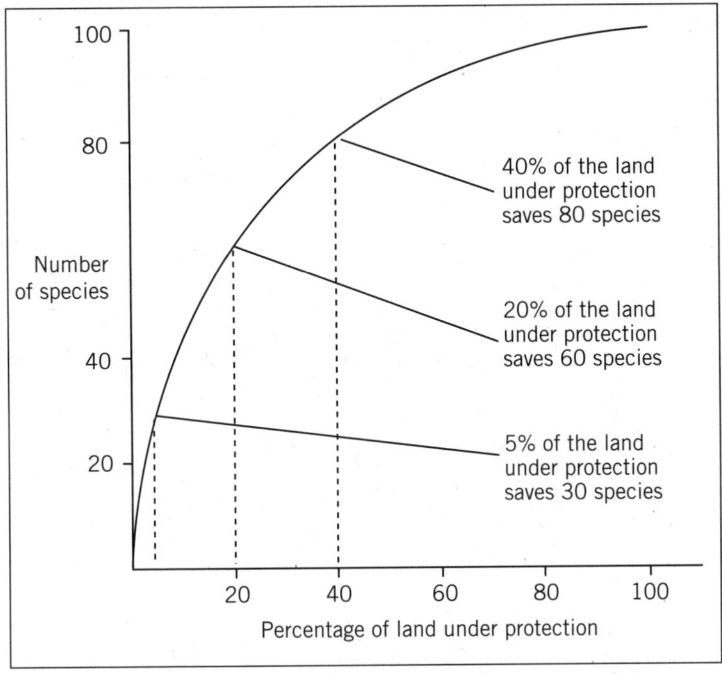

Figure 9.6. Biodiversity (number of species) preserved within protected areas as the percentage of land under protection increases.

also on the basis of its impact on the biodiversity of the areas in which it is to be introduced. Just as zoning laws sometimes prohibit particular activities in certain areas of cities, zoning laws in potential agricultural and forestry areas ought to be enacted. One consideration for these laws should be the impact of any activity on biodiversity.

While we do not imagine that such laws will be enacted in the near future, we hope that as consciousness is raised concerning the loss of biodiversity in forestry and agricultural areas within rain forest habitats, movements in this direction will evolve. We return to this political point below, but first let's explore what this alternative model for biodiversity preservation might look like.

In Figure 9.6, we illustrate a typical graph of the number of species plotted against area. The area may be just the size of a single small sample in a region, or it may be the size of a preserve or group of preserves. The number of species will tend to rise as the area increases, but eventually level off at some specific value. We have illustrated a situation in which there are a total of 100 species in the entire region.

Note how the number of species we can expect to be preserved drops as the area devoted to preserves decreases. With 40 percent of the land under reserve status, we save 80 of the 100 species. With 20 percent in reserves, we save only 60 species, and the number of species saved drops to 30 if we preserve only 5 percent of the land. This is a well-known relationship—the number of species increasing with increasing area—and has led many to suggest that for maximum conservation of biodiversity the size of biological preserves must be as large as possible. This seems so obvious. But is it?

If we back up and ask about the land outside of preserves, and take that land into account in our calculations, the picture begins to look quite different. If we assume, as many conservationists are willing to do, that close to 100 percent of the species will go extinct outside of the reserves as modern agriculture continues its assault on nature, then the analysis is true—the larger the reserve, the better. But if we concern ourselves with the management of the land outside of the reserves, it is not at all clear what the correct strategy might be. For example, if we put into place regulations that require no more than a 50 percent loss in species diversity in those areas under management, putting 20 percent of the land in reserves will save an equal number of species as would putting 40 percent in reserves (the sixty species preserved in the reserve plus twenty of the remaining forty species that do not go extinct because of the regulation). We display, for the example in Figure 9.6, the expected biodiversity preservation as a function of the various management strategies in Table 9.2.

Table 9.2. Hypothetical biodiversity preservation (out of a total of 100 original species) under various management strategies.

Size of reserve as percent of total area	**Allowable extinction outside of reserves**		
	10%	**50%**	**100%**
40%	98	90	80
20%	94	80	60
5%	84	65	30

This table makes it clear that an overall strategy that includes concern with what happens outside of the biological reserves is more likely to conserve the greatest biodiversity. In the real world it is not likely (nor is it desirable) that biological reserves will encompass all of a region, and the most that can be expected is something more on the order of 5 percent of the land in reserves (Costa Rica, for example, is one of the world's largest preservers, with 10 percent of the national territory in strict preserves). But even with this small quantity, a great deal of biodiversity will be preserved if we simply concern ourselves with the loss of biodiversity in the managed ecosystems outside of the reserves. In the above example, with only 5 percent of the land under formal reserve status, but with the requirement that management activities not reduce biodiversity by more than 10 percent, more species are saved than in the case of 40 percent preserved in pristine islands amidst a sea of devastation.

Unfortunately we know too little of biodiversity in general, or of biodiversity in managed ecosystems like agriculture and forestry specifically, to be able to judge positively which model, the classical or the alternative model, will preserve more biodiversity. But given preliminary studies, and given the poor record of modern agriculture in biodiversity conservation, we strongly suspect that our alternative model is more appropriate. In the shaded coffee agroecosystem, for example, biodiversity of arthropods appears to be on the order of 60 to 70 percent of what the original forest might have held. Converting that agroecosystem to its highly capitalized modern form reduces that biodiversity to perhaps 5 percent of the original.

The alternative argument is sometimes made that the highly capitalized modern agroecosystem is sufficiently productive on a per-hectare basis that a smaller amount of land can be put into agricultural production in order to satisfy human demand, thus leaving the rest for biodiversity conservation. Putting aside the abominable record of "modern" agriculture in terms of satisfying human needs,[19] we must ask how much biodiversity is lost in the transformation to the modern state, and how much is preserved by keeping the rest in a pristine state. This is not a simple issue, and indeed we know too little about biodiversity in managed ecosystems to give a proper assessment. But again, we strongly suspect that a terrain of biological devastation dotted with islands of pristine rain forest will preserve far less biodiversity than a landscape of sustainable agricultural and forestry activities surrounding a smaller number of pristine areas.

Furthermore, a recent study[20] examined increasing agricultural production efficiency as it relates to biodiversity preservation. Before we examine the results of the study, we will note that the common assumption that higher productive efficiency in agricultural areas will reduce pressure to cut more forest is at odds with standard economic development theory. Standard development theory holds that as production efficiency at one locality increases, humans and capital will tend to migrate to that locality. Thus, standard economic theory would suggest that as production efficiency increases at one point, more people will be attracted to that point, and consequently more land for agriculture will be required, which will mostly be taken from extant forested areas. The study in question considered nineteen cases, asking whether the standard economic theory or the conservationists' theory more frequently fits the data. With rare exceptions, the standard economic theory seems to explain the data better than the conservationists' version; that is, a locally elevated productivity through technological improvement tends to attract migrants to the area, which tends to encourage further deforestation. Thus, the dream of increasing agricultural production efficiency so as to require less land to produce seems to have the opposite effect to that expected by traditional conservationists, further supporting the model we propose as an alternative.

However appealing our alternative model might seem, political realities are something different. Consider again the coffee agroecosystem. In the early 1990s the price of coffee bottomed out, mainly due to post–cold war political maneuverings. The problem was overproduction on a worldwide scale, a fact agreed upon by economists, planners, and coffee farmers alike; but coffee agronomists continued to insist on an almost religious commitment to increasing production—what has sometimes been referred to as the "productionist mentality." Most coffee producing countries formed a new informal cartel in 1992, and prices began rising in 1993 as a result of the members' agreeing to retain significant quantities of coffee off the market.[21] This case presents a dual problem: a biodiversity crisis in coffee-producing areas; and an overproduction of coffee. If it is true that a significant amount of general biodiversity is sequestered in traditional coffee plantations, it would make sense to try to preserve them, which means preserving the traditional ways of coffee production. But classically trained agronomists argue that the traditional ways are not very productive—that is, the amount of coffee produced per unit of area is much lower than in the modern system (a debatable claim at best). Yet since all agree that the major problem with coffee is overproduction, why should we value producing more per hectare?[22] Perhaps we shouldn't, but in the current system we do—to everyone's detriment. Individual farmers (or nations) try to produce as much as possible, but in so doing, they collectively saturate the market, and prices are consequently reduced—a tragedy for all.

An ideal solution in this case might be to give "biodiversity credits" to those farmers who decide to return to traditional production, instead of paying farmers to keep coffee off the market. Investing government funds in this sort of conservation would treat two problems simultaneously: the loss of biodiversity through agricultural modernization and the overproduction crisis in coffee.

Still, the argument is frequently made that putting only a few farms in high yield production and letting the rest revert to forest may also preserve more species and reduce production so as to raise prices. We repeat: not enough is known about the patterns of biodiversity in agroecosystems to say which of these two strategies would preserve the most biodiversity, although the record thus far does not bode well for the high-yield strategy. Our prejudice is that an entire landscape devoted to sustainable organic coffee production, using shade trees, fruit trees, plantains, and understory crops as part of the system, is likely to preserve more biodiversity than islands of forest in a sea of modern pesticide-drenched agriculture.[23] And, to repeat a theme of previous chapters, who will benefit from a few highly technological coffee farms? And what will the rest of the people do for a living?

It is on this note that we must end our discussion of biodiversity per se. In Chapter 8 we already devoted considerable space to the landscape model we propose. We can only add, recalling our earlier argument, that even if it turns out that biodiversity is *not* more completely preserved in our landscape model than in the high-productivity model, there are social and political reasons why ours is the preferable—perhaps even the only viable—model. But we must emphasize: our landscape model is simply not possible without land security.

10: WHO CONSTRUCTS THE RAIN FOREST?

IN A BEAUTIFUL rain forest preserve in the Caribbean lowlands of Costa Rica, one of our students remarked, "The trouble with this place is that even on the back of the property you can hear trucks on the road." You might ask what trucks or the absence of their sound have to do with a tropical rain forest. The answer is nothing and everything. To this student, a certain feature of the tropical rain forest was the solitude, the isolation, the feeling of being removed from the modern world. His point of view is shared by many who have spent time appreciating the natural world. Nature, in this view, is what was there before human systems came to dominate and alter it. Furthermore, nature is always seen as more beautiful than its human-modified form. This idea is a social formation that we call nature worship.[1]

A radically different idea of the rain forest is suggested by the Ston Forestal company. Ston maintains huge monocultural plantations of melina, a nonnative tree that grows very rapidly. To take advantage of substantial tax breaks offered by the Costa Rican government for "reforestation," Ston does not call its operation a tree plantation, but rather a reforestation effort. The result of their effort is, by legal definition, a forest. This conceptualization of a rain forest is probably dictated more by the desire for tax credits than by any heartfelt feeling about tropical rain forests.

Looking back in history, we see similarly dueling visions of the rain forest. Consider the following description of a tropical rain forest:

> . . . broken only by the clanging clamor of the torrents, the growling of tigers, and the swarming of infinite vipers and venomous insects . . . the plague of vampire bats . . . extends until Brazil, treacherously sucking in the hours of dreams the blood of men and animals. There also, by the side of the Brosymum [tree] grows the Rhus juglande . . . whose mere shade puffs up and scars the careless wanderer. There one begins to suffer the privations and calamities of the wilderness, that . . . grow so as to at times make of life there scenes whose horror could figure in the Dantean pages of Purgatory and Hell.[2]

A more recent assessment suggests that the tropical rain forest is

> . . . in a class of its own. It is akin to other super-size spectacles on Earth, such as the Grand Canyon and the Victoria Falls. However much you read about the scene, however many photographs or films you see of it, nothing prepares you for the phenomenon itself with its sheer scale and impact. You gaze on it and you feel your life has started on a new phase. Things will not be the same for you again after setting eyes on something that exceeds all your previous experience.[3]

So what, in point of fact, is a tropical rain forest? Is it a hell or a paradise? Is it a place to clear for a farm or to look for birds? Is it somewhere to escape from the modern world or a plantation of melina trees? Is it something that contains cures for cancer, or is it an impediment to cattle and a source of poisonous snakes, agricultural pests, or enemies of agricultural pests? Clearly there are at least two components of the tropical forest: (1) its trees and other plants, microbes in the soil, and insects and other animals, in their many relations to one another; and (2) the observer, who may be a local farmer, a foreign industrialist, an ecotourist, or someone else entirely. In short, there are a text (1) and a reader (2). We believe the nature of the tropical forest can best be understood as the relationship between the two.

Our use of the terms "text" and "reader" is intended to call attention to the resonance between recent theories of literary criticism and the notions of nature that we humans have. Contemporary literary theorists tend to view texts within the context of who is reading them. It is impossible, they argue, to analyze a text independently of its context, and its most important immediate context is the person engaged in reading it. True meaning, then, is not to be found in the text alone, but rather in the *reading* of the text—the interplay of the actual words and the context in the reader's mind.[4]

If the text, so to speak, is the collection of physical elements in the area called forest, and the reader is the person who observes that text, then the dramatically divergent views of the tropical forest with which we opened this chapter are easily reconcilable. For our student, the forest was a pristine wilderness. In reading the text, what stood out in his mind was the sound of the trucks on the nearby road. For the CEO of Ston Forestal the forest is read as a bunch of trees waiting to be harvested for profit. The Spanish explorer, searching for gold mines, read the text as akin to "Dantean pages of Purgatory and Hell." The contemporary first world conservationist reads the forest as something that makes your life start on a new phase. Thus the nature of the tropical forest is as much a social construction as the Pyramid of the Sun in Mexico, the Holy Trinity, or the United States' peculiar notions of democracy.

If a tropical rain forest is as much a social construction as other human structures, both physically and conceptually, it may be "deconstructed" and "reconstructed." Is this not what our student did when he suggested that a true tropical rain forest is not simply the plants and animals that make up the physical aspects of the forest, but includes the lack of truck sounds? Did not Ston Forestal deconstruct the romantic notions of what forests are and reconstruct, largely for the purpose of tax credits, the forest as nothing but a stand of trees? And the Spanish explorer and conservationist likewise have constructed their own notions of the rain forest. If the nature of the rain forest is constructed by our reading it as a text, we must ask: whose reading of the rain forest represents truth?

This sort of reasoning can easily lead to a quagmire of excessive relativism—if everyone can have a different reading of the same text, there is no true meaning. It is most definitely not our intention to promote such a view. However, we do feel it is imperative that we deconstruct our notions of nature as a means of probing the social relations that give the concept meaning in the first place so as to develop a more useful social construct of the rain forest as a part of that nature.

If the nature of the rain forest is to be adjudicated as a social construction (as we argue it must), part of the evaluation of any particular construction ought to question who is doing the constructing in the first place. And in the present world, there are many actors who do that construction. The essence of the rainforest is thus not only a combination of text and reader, but a negotiated conceptualization involving a variety of readers, each of whom, necessarily, has only a partial reading—the Spanish explorer, the first world conservationist, the peasant farmer, the plantation owner. And we can fully expect that conceptualizations of the rain forest will change as different readers engage with the text and interact with one another.

Our own constructions of nature are complex. John's construction of nature arises from (among other things) landscape paintings, boy scoutish nature lore, summer walks in the north woods with his mother, fishing with his father, adventure movies like *Tarzan* and *King Solomon's Mines,* teaching ecology at the university, personal relationships with others who love the outdoors, sharing visions of pristine wilderness with friends and acquaintances, and so on. Ivette's construction arises from (among other things) weekend excursions to the El Yunque rain forest, the beaches of Guánica with their pollution from petrochemical companies, television nature shows, a fascination with the mystery of the ocean and its fauna, visits to her uncle's farm, lizards on the walls of her house, teaching tropical ecology at the university, and more. Obviously both are unique visions, despite their feeling of generality. It is difficult for each of us to imagine that our view may not be the "correct" construction. But if our previous argument holds, there is no clear way to prioritize our construction over the construction of the peasant who must cut down the forest to feed her family. It gives the economist at Ston Forestal equal reason for arguing his view as it does our student for arguing his, a disturbing idea to be sure. Despite such apparent relativism, there remains great utility to the idea that the rain forest is a social construction—which is to say, a set of physical features (the metaphorical text) combined with human observers of those physical features (the readers).

The utility of this point of view—the idea of nature construction—may be appreciated by an analysis of power structures in society. If nature is socially constructed, it seems likely that similar sociopolitical conditions will produce similar constructions of nature. Capitalists seek to construct nature in a way that will maximize profits. Thus, a poisoned river is not necessarily a negative in the capitalist's construction. Peasant farmers seek to construct nature in a way that will allow them to survive. Thus, a deforested patch of land is not necessarily a negative in the peasant's construction. Workers seek to construct nature to preserve or improve employment conditions. Thus, global warming is not necessarily a negative in the worker's construction. The content of one's construction of nature is based in part on one's power position in society.

Furthermore, all constructions will necessarily be partial since all experience is partial. As political and other alliances form, aggregate constructions will change. Those alliances that achieve a share of political power will contribute more to a society's dominant construction of nature than those alliances that fail. Thus a society's aggregate construction of nature will always be a partial one based on shifting political alliances. The particular forms of intended nature conservation will emerge from those political alliances.

Strategies for Rain Forest Conservation

In the same sense that one cannot envision the exact nature of an egalitarian society, yet can actively engage in the struggle to attain such a society, one may not see exactly how to construct the nature that is to be preserved, yet can actively engage in the struggle for a progressive conservation of nature. The theoretical principle of egalitarianism is clear enough, and the theoretical principle of conservation is clear in the same sense. However, just as the "egalitarian" Athenian democracy, which was based on slavery, is hardly regarded as egalitarian today, the "conservation" of today's conservationists may be unrelated to a future construction (we think it likely). The construction of nature as created by today's educated classes—including parts of our (John and Ivette's) own constructions—may bear little resemblance to constructions to be elaborated in the future. And there is little question that future constructions will be formulated by different political alliances than those that exist today. Indeed, since we first wrote this book, the idea of a landscape mosaic has become far more commonly accepted as an obvious conservation goal than it ever had been before—clearly a change in the social construction of conservation.

Our social construction of the rain forest is distinct from those of a local peasant farmer, a banana company executive, or an indigenous person living in the forest. We believe that as (and if) society evolves into a more egalitarian form, some of these distinctions will tend to diminish. Other differences may emerge, but those based on class will necessarily disappear as classes disappear. When the experiences of all people are constrained by similar material forces, certain social constructions of nature will likely be far less variable than they are currently. Most would agree that the utopian future of an egalitarian society is (alas) hardly on the horizon. Yet some speculation on such a future may serve to initiate the debate on what might eventually constitute a "correct" construction of nature. In that spirit we offer a very brief speculation here.

The construction of nature that sees much of the natural world as being worthy of preservation, reconstruction, rational use, and so on has become popular in the West due to the thoughts and writings of privileged classes of people, mainly from the developed world. It reverberates in so many ways with the reported constructions of many indigenous people, at least those less tainted by the ideological assault of western capitalism. Yet it is frequently at odds with the constructions of the masses of poor peasants and slum dwellers of today's Global South. The poor who struggle to make ends meet in the global system have little time to contemplate the merits of biodiversity preservation when their daily survival is at risk—though their ancestors may have shared a spiritual connection with the natural world similar to those of contemporary western conservationists.

Apart from utilitarian concerns that saving nature will find new cures for cancer, or help stop global warming, western conservationists from the privileged classes may honestly argue that they wish to save the rain forest (or coral reef, prairie, or desert) because they find it immensely aesthetically pleasing; that cutting down rain forests would be like whitewashing the frescoes of Diego Rivera. Utilitarian claims notwithstanding, we believe this argument is the motivation for the vast majority of first world conservationists, and it has a strong affinity with the spiritual values of many indigenous people. If these particular readings of the text of nature are of value (and we believe they are), it makes sense to promote them. And the best way to promote a particular reading is to afford all potential readers the same reading privileges. But the peasant farmer struggling to feed his family is not likely to have those privileges. Thus, the social construction of a nature that is beautiful and worth saving is not likely without the reorganization of contemporary structures of political power. In short, it will require the elimination of the situation in which some people are forced, by their class position, to read a negative text as they participate in the social construction of nature—to see the rain forest as something to be eliminated, not venerated. If we feel that a view of nature as something to be appreciated is even partially correct, our promotion of that view needs to take the form of a struggle for social justice.

The "Backlash" Strategy

Subsequent to the first edition of this book, several books were published with quite a different perspective than ours. Two highly acclaimed examples are John Terborgh's *Requiem for Nature* (1999) and John Oates's *Myth and Reality in the Rain Forest: How Conservation Strategies Are Failing in West Africa* (1999). While they received some harsh criticism in the technical literature,[5] they were generally met with rave reviews in popular media and were frequently cited as works willing to tell the hard truth to an idealistic yet naive audience of do-gooder conservationists. While we agree with a great deal of the practical analysis that went into the construction of their arguments, we feel that in the end they missed the point.

The basic argument of these so-called backlashers is that conservation programs that have emphasized the needs of local people have not worked. Despite sincere efforts by conservation organizations to take the needs and desires of local people into account, the trashing of tropical rain forests continues unabated. These authors generally criticize case after case of conservation programs designed to bring local people into the management scheme or provide local people with some economic benefits from the conservation efforts. None of these programs has worked, they argue, and the entire effort to involve local people in conservation efforts should be abandoned. While we would love to see their logic applied in other cases (i.e., the drug war in the U.S. or the structural adjustment programs of the International Monetary Fund, to take two examples), the simple argument that something should be abandoned only because it has not yet worked seems a little silly when viewed historically. Yet we largely agree with these authors' overall critique in the sense that most such programs have, as they say, been utter failures. Where we disagree is with their programmatic conclusion—that the only solution is to set up isolated reserves of highly protected natural habitat, free of human influence.

The chord of truth struck by these analysts is that almost all conservation programs designed to decrease the destruction of tropical rain forests have been embarrassing failures. What the backlashers fail to recognize is the obvious (but critical) fact that these programs have failed whether or not they incorporated the local-people factor. For example, the adulated Costa Rican program of the 1980s was a stark failure that involved no local people; currently, much of Costa Rica's reforestation has little to do with conservation efforts. So while we can agree that conservation programs aimed at preserving tropical rain forests have been failures, we fail to see any concrete evidence that the reason for those failures is the inclusion of human welfare in their formulation.

Let us remember that the problem under analysis is the destruction of tropical rain forests, which is a sociopolitical issue. The strategies of the conservationists criticized by the backlashers were born of the realization that razor wire fences and armed guards will not protect rain forests in perpetuity. Those whose dream is the wilderness equivalent of the gated community need to understand that wilderness destruction is a sociopolitical problem requiring a sociopolitical solution. There is certainly much room for discussion as to exactly what that sociopolitical solution ought to be, but the response to "put a fence around it" is not likely to work, except for short periods of time in a very few places. Authoritarian governments might force such a strategy to work temporarily, but when they fall (as they invariably do), what happens to the razor wire and the armed guards? Band-aid solutions, such as those that provide desperately poor people with the hope of being maids in a foreigner's ecotourist hotel are not likely to work either (as the backlashers correctly point out). We feel that both these approaches—armed exclusion or feeble "inclusion" of local poor people—have been discredited by experience. Rather, it is essential to analyze the web of causality, look for its weak points, and do more than just tweak its strands. If the web itself is the cause of ongoing rain forest destruction, the web must be broken. The question is not whether a band-aid solution will work. The question is how to break the strength of the web that holds together the complex sociopolitical system that causes the forest to come down.

Perhaps the backlashers got it biologically wrong also. As we noted in Chapter 2 and Chapter 9, recent ecological research strongly indicates that protecting isolated patches of forest may not be sufficient to avoid massive extinctions. The technical arena in which this sort of problem is discussed is called "metapopulation dynamics," and it basically says that living beings must be able to recolonize areas where they have locally disappeared, if long-term extinction of the entire species is to be avoided. In this analysis the long-term prospects of isolated preserves seems less optimistic than originally conceived by conservationists. What appears to be the case is that landscape planning that ensures avenues of dispersal between fragments of natural habitat is essential for long-term conservation of species. Thus, even if we ignore the sociopolitical side of the argument, this research suggests that the "wire fence/armed guard" approach is also ecologically flawed.

Social Justice and Rain Forests

The image of peasant farmers cutting down rain forests has been a potent one in the past. In a very crude form it is the image that drives people to a variety of false analyses, such as the neo-Malthusian theory that overpopulation drives destruction. Indeed, in the example of the Sarapiquí, we have heard respected scientists claiming that the problem in the area was simply the one of too high a birth rate amongst the local peasants. Our position is clearly at odds with this point of view. Yet we too gain certain inspiration from our interactions with and observations of peasant farmers. While the assumption that convincing peasants to have fewer children will help with the problem of deforestation is misguided, the observation that landless peasants do in fact cut rain forests down is accurate. We share a certain subtle conviction with the neo-Malthusians in that we feel there are "too many" poor, landless peasant farmers. Our solution to the problem is, however, quite different.

We have gone to great pains to demonstrate that peasant farming activity, while directly responsible for much deforestation, is itself a consequence of something else. Thus, to engage in activities aimed at stopping deforestation by peasants, it is necessary to aim at the causes of the deforestation, which are fundamentally structural. The politico-economic structures that define the way the Global South functions are the structures that inexorably give rise to the peasantry to start with. The question of why the rain forest comes down may have the proximate answer, to make way for peasant agriculture. But one must be studiously myopic not to move right to the next question, why are there landless peasants who need to cut forest? This does not have as simple an answer, yet that answer is the key to intervention in the problem of deforestation.

Furthermore, to ask why there are landless peasants is much the same as asking why there is poverty—which is the fundamental question of progressive politics at its most basic level. The movement for social justice has represented a foundational challenge to inequitable sociopolitical structures as far back as history records. And here we see the connection. If Why are there landless peasants? is the key question for deterring the destruction of rain forests, and also the foundational question for the politics of social justice, then the movement to save rain forests needs be closely linked with—if not take on virtually the same form as—the movement for social justice.

11: PAST CAUSES, FUTURE MODELS, PRESENT ACTION

THE THEME OF THIS BOOK has been, in various guises, the causes and consequences of rain forest destruction. Implicit in this theme has been the subtheme that an alternative model of socioeconomic organization could drastically change those causes. Furthermore, certain programmatic conclusions seem to emerge from this analysis. In short, the book has sought to pose and answer three fundamental questions:

1. **Question:** What causes rain forest destruction?
   Answer: There is a web of causality, no single component of which is truly the cause.
2. **Question:** What is a model for the future?
   Answer: A planned mosaic, based on ecological and egalitarian principles.
3. **Question:** What is the political action plan?
   Answer: Intensify the struggle for social justice.

The Web of Causality

The notion of the web of causality is key to understanding why we face the problem of rain forest destruction in the first place. That has been a theme throughout this book. We illustrate the basic features of that web in Figure

11.1. Reading from the bottom of the figure, we see that damaged rain forests will recuperate if not further damaged, but will recuperate far more slowly if further modified by agriculture. The damage itself is created by either logging or modern agriculture, and the damaged forests are further cleared by peasant farmers. But the peasant farmers' activities are a consequence of the opportunities created by logging as well as the ups and downs of the international market, which cause modern agriculture to hire and fire workers. Modern agriculture needs those workers, as well as the land that it buys or steals from peasant farmers. It is quite pointless to try to identify a single entity within the web of causality as the "true" cause. The true cause is the web itself.

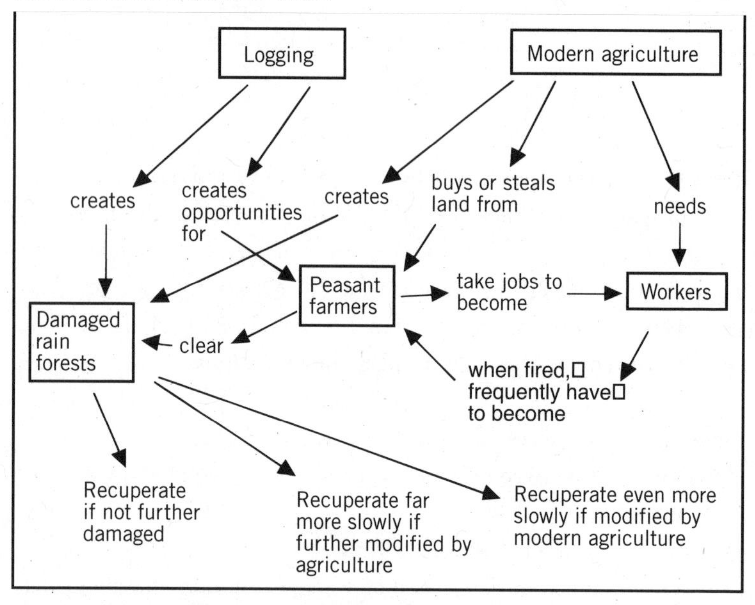

Figure 11.1. The web of causality for deforestation of rain forests.

Yet even this is an oversimplified picture. The web of causality is not just the farmers, loggers, modern agriculture, and workers. That is just a subweb embedded in a larger web, as illustrated in Figure 11.2. With the picture of the expanded web, the international nature of the problem becomes apparent: the international banking system, national governments, the U.S. and other developed world governments, and consumers and investors in the developed world all play key roles. This is the true web of causality, and it is a complicated, tightly woven one. Tweaking one strand is not likely to bring the whole thing down. Fighting a concerted battle to restructure the entire nature of the web will.

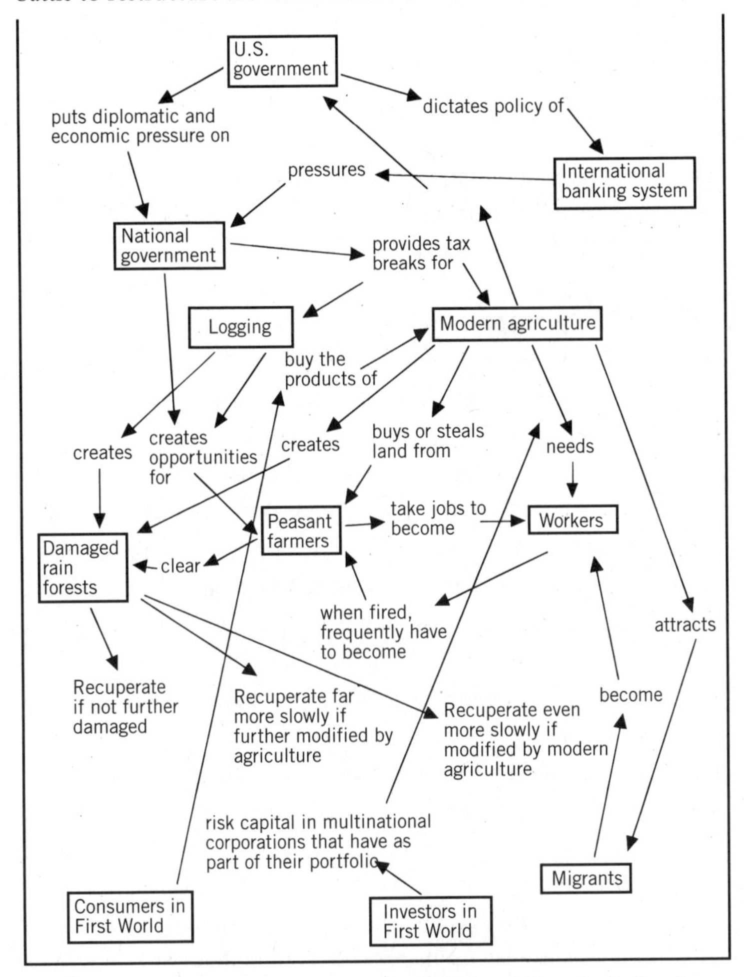

Figure 11.2. The expanded web of causality for deforestation of rain forests.

Furthermore, seeing the entire web of causality enables highly focused political actors to see their relation to other actors, and enables an analysis of consequences that may be dramatic, even if indirect. For example, organizing consumer boycotts can be seen as clearly attacking the arrow that connects the consumers to modern agriculture through the arrow "buys the products of." But following through the logic of the web also suggests that a successful consumer boycott may likewise affect the arrow from modern agriculture that "needs" workers, thus creating more peasant farmers, who will likely clear more forest. If a careful analysis of this situation reveals that the loss of jobs will be severe, the political action agenda of the boycott might then be expanded to form alliances with a local farm worker union calling for job security or a political movement that seeks secure land ownership for the increased number of peasant farmers that will surely be created if the boycott is a success.

The Planned Mosaic

In thinking of a model for the future, we are struck with two images from Central America. Figure 11.3 is a photograph of a poster that was on the wall of CORBANA, the technical research organization of the banana industry in Costa Rica. It is presented as a utopian future in which agricultural fields will be virtually reconstructed from their original state, crisscrossed with irrigation channels and subtended by plastic drainage tubes, all automatically installed with sophisticated laser-guided machinery. Nature has been fully tamed, down to the physical reconstruction of the water table under the soil. And in the model of isolated islands of pristine rain forest surrounded by biological deserts of pesticide-drenched modern agriculture, this is the utopian dream of the modern technocrat. It is also a utopia that fits well with the "fence-out" approach to conservation.

The other image is presented in Figure 11.4. This is a photograph of a poster that was hanging in the office of the Ministry of Agriculture of Nicaragua in 1988, during the Sandinista administration. It illustrates a diverse mosaic of land uses, ranging from protected forests to managed forests to plantations to sustainable agriculture, a mosaic where decisions about land use are tied to the capabilities of the land and the needs of people, not to the requirements of profit or repayment of past accumulated debt, or even the desires of northern conservationists. This is quite a different vision. Nature has not been tamed, but is something of a partner with the humans that live in it. This vision is the one that has a far greater

Figure 11.3. Poster on the wall of CORBANA, the technical research institute of the banana companies in Costa Rica, emphasizing the technical mastery of nature.

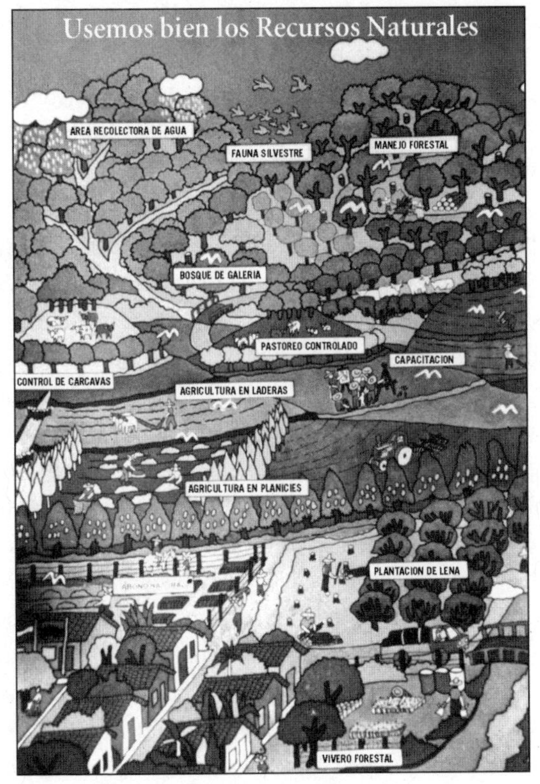

Figure 11.4. Poster on the wall of the Ministry of Agriculture in Nicaragua in 1988, emphasizing the ecological mosaic of land uses.

chance of reordering that web of causality and stopping, perhaps even reversing, the tide of deforestation.

The Political Action Plan

All this is meaningless without a program of political action. While we don't intend this book to be specifically about political action, certain principles do emerge from our analysis of the problems. Political action must focus on the web of causality and eschew single issue foci. Calls for boycotts of tropical timbers or bananas need to be coupled with actions to change investment patterns and international banking pressures. Above all, political action plans must be formulated so they at least do not make the situation worse, which is certainly conceivable given the complex nature of the web of causality. But as the various strands of this web are brought into the analysis, it becomes obvious that political action needs to be focused not only on rain forests and the subjects traditionally associated with them, but also on the same sorts of issues that have motivated progressive forces in the past: issues of social justice. The same peasant farmers that formed the backbone of the Vietnamese liberation armies or the Salvadoran guerrilla forces are the ones forced into the marginal existence that compels them to continually move into the forests. So the issues that have in the past compelled progressive organizers to form solidarity committees and antiwar protests are the same issues that must be addressed if the destruction of rain forests is to be stopped.

Furthermore, just as the most effective political action in the past was organized in conjunction with and to some extent under the leadership of the people for whom social justice was being sought, so today political action should be coordinated with those same people. As that coordination proceeds, the alliances that form will inevitably lead to the reformulation of goals, which the rain forest conservation activist must acknowledge and be willing to respond to. The local people, quite obviously, must recognize that it is in their best interest to conserve rain forests, and it is the job of the progressive organizer to construct the political action so that such value is evident. In short, the alliance between the people that live in and around the rain forest and those from the outside that seek to stop the tide of rain forest destruction must be a two-way alliance. If the people who live around the Lacandón forest in Mexico, for example, have as their major goal the reformulation of the Mexican political system, the rain forest conservationist must join the political movement to change that system, something that many would see as distant from the original goal of preserving rain forests. Political action to preserve rain forests, under the framework of the web of causality, will inevitably involve the serious activist in social justice issues that initially seem only marginally associated with the problem of rain forest destruction.

This political action plan takes on far greater significance today than it did when we first presented it in 1995. The growth of the international globalization movement, as described in Chapter 7, has in a sense put into practice what we had advocated. Issues of biodiversity and rain forest protection have become part and parcel of the overall movement, included in it as naturally as workers' rights and democracy. Today, the political action plan based on the web of causality has become more than just a theory.

The web of causality, the landscape mosaic, and the social justice political action framework thus form the tripod upon which we seek to base the conclusions of this book. In the end we recall our initial image of slicing bananas on our cereal in the morning fading into the logger's saw slicing the trunks of the rain forest. We intend for this imagery to continually recall the interrelationship among various activities in the web of causality, and to fix the image of banana workers forced to work in substandard conditions before they get fired and are forced to look for a piece of land they can farm—sometimes cutting a patch of rain forest for that purpose. A breakfast with sliced bananas thus becomes a metaphorical breakfast of biodiversity. And as integral parts of the same web of causality, slicing the bananas and cutting the trees are, in the end, more than just components of a metaphor.

We close this edition of *Breakfast of Biodiversity* with the story of Marcelino. We have been involved in a research project on the Caribbean coast of Nicaragua for the past fifteen years, visiting the same communities once a year, and have come to know some of the people there quite well. One young man, Marcelino, had been a constant companion of ours during much of that time. Marcelino was thirteen years old the last time we saw him. He regularly accompanied us to our rain forest study plots every year. He was a delight to have around. He was extremely inquisitive and had picked up on a lot of rain forest lore from us—as much as he provided us with local knowledge of the same.

After some five years of Marcelino's hanging around with us, his mother told us that because of his time with us he has come to love the forest and now has a dream of going to school to "learn how to study forests." When we heard this, we were at first overjoyed at the apparent effect we had on this local kid. We had made a new conservationist! A local boy would spread the word of the beauty of the forests and the need to preserve them. But only a few minutes of serious reflection reminded us of the realities of the web of causality. The truth is that Marcelino's fifteen-year-old brother had already migrated to Costa Rica to work in the banana plantations. Marcelino's family has only a small plot of land, not really enough to feed the family even in productive years. If he stays, he faces a future of trying to hire himself out as a machete wielder, cutting weeds in someone else's plots. He is extremely intelligent and would probably become quite frustrated with this life. Costa Rica will look very enticing—especially since his brother has already found work in the banana plantations—and we fully expect he will also migrate.

But the life of a banana worker is hardly stable, and he will almost certainly find himself out of work periodically. He will also likely start a family. Given his drive and intelligence, he will surely want to make something for his family, and an insecure job in a banana factory will not likely be satisfying for him. But what else is there for him? Go back to Nicaragua where there are no jobs at all? Move to the shantytowns near the city? Probably not, since he has been a rural resident all his life and is practically illiterate. Most likely he will be forced to do what many others have done: find a piece of forest, cut it down, and make a small homestead for his family. Thus, we begin with a thirteen-year-old boy who loves the forests and wants to learn how to study them. And circumstances will make of him a twenty-year-old man cutting down those very forests. The web of causality spins its inevitable structure, capturing this powerless individual, and the forest comes down.

NOTES

Chapter 1

[1] While Costa Rica's short-term debt actually decreased from $575 million in 1980 to $341 million in 1992, probably due to the particular political situation of the 1980s, as described later, its long-term debt actually increased from $2 billion to $3.5 billion during that same time period, an increase of 68 percent. This is not as bad as some other Latin American countries. El Salvador, for example, increased its debt by 208 percent during that same period, and Mexico managed a 101 percent increase. Compared to other Latin American countries, Costa Rica is doing better today than it was in 1980, but is still one of the most debt laden of all—worse than Mexico. Only Panama, Argentina, Venezuela, and Nicaragua are worse off.

[2] A hectare is equivalent to 2.47 acres.

[3] The La Selva Research Station, Braulio Carrillo National Park, Tortugero National Park, and the corridor between La Selva and Braulio Carrillo.

[4] Butterfield, 1994; Montagnini, 1994.

[5] Thrupp, 1991.

[6] Most of the history in this section comes from conversations with the late Raphael Echeveria, a long-time local resident and former banana worker and timber cruiser. Information is also derived from Danilo Brenes and Hector Gonzales. The history is effectively the same as presented by Butterfield, 1994.

[7] A general introduction to worldwide patterns can be found in Tucker, 2000.

[8] Lewis, 1992.

[9] In 1989 the president of the Solidarista organization at the Standard Fruit Co. plantations in Rio Frio proudly told one of us (Vandermeer) that "anyone trying to organize a union will be fired. But we know who most of them are and we won't let them get jobs here in the first place" (paraphrased).

[10] A great debate has arisen about the idea of a pristine rain forest. Most areas formerly thought to be uninfluenced by the hand of *Homo sapiens* turn out to have been occupied by prehistoric populations.

Chapter 2

[1] Majer, 1980.

[2] Vandermeer, et al., 1995.

[3] Janzen and Schoener, 1968.

[4] Höldobler and Wilson, 1990.

[5] Terborgh, 1992.
[6] Huston, 1995.
[7] Werner, 1992; Schoener, 1993.
[8] Yih, et al., 1991; Mooney and Gordon, 1983
[9] This basic idea is known as "recruitment limitation" and is discussed fully in Hubbell, 2000.
[10] Denslow, 1987.
[11] Phillips and Gentry, 1994.
[12] Richter and Babbar, 1991.
[13] However, as discussed in the next chapter, if there is a general characteristic that applies to all tropical soils, it is their heterogeneity. Having most of their experience in temperate regions of the world, many soil scientists have viewed tropical soils as uniformly acid and infertile. More recent studies have emphasized that this view is incorrect (Sanchez, 1976; Richter and Babbar, 1991)—that soils vary from highly fertile to highly infertile in the tropics. On the other hand, for most regions of the tropics, agricultural conversion has already discovered the good soils, as discussed in Chapter 3, and the remaining rain forests are generally on very poor acid soils.
[14] The species of this family found in South America is a recently described species that is placed in its own subfamily, but its relationship with the Asian and African dipterocarps is somewhat enigmatic (Mabberley, 1992).

Chapter 3

[1] There is some debate about the question of organic matter in rain forest soils. The rate of decomposition of organic matter is about twice the rate it is in a normal temperate zone situation, and thus it is only natural to expect the standing crop of organic matter to be less in the rain forest. Some authors have questioned this basic assumption (e.g., National Research Council, 1993). On the other hand, all are in agreement that once the forest is cleared for agriculture, whatever organic matter was actually there rapidly disappears from the soil.
[2] For anyone interested, the USDA system would classify the acid soils as either ultisols or oxisols, the alluvial soils as either inceptisols or entisols, the volcanic soils as andisols, the hillside soils under a variety of categories, and the swamp soils as histosols (USDA, 1975).
[3] This point was discussed in Chapter 2.
[4] "Clay particle" in this sense is a crystal chemical molecule. The molecule itself becomes exceedingly large since it is built up on the basis of a repeating system of strong chemical bonds.
[5] Remember that pH is simply the proportion of positive to negative

charges in the solution, with greater positive charges leading to greater acidity and vice versa.

[6] The red color of many tropical soils is due to this excessive amount of iron.

[7] Recall that humus, or decaying organic matter, acts physically like clay particles, with net negative charges on its surface, as discussed in Chapter 2.

[8] Westoby, 1989.

[9] Shiva, 1993.

[10] Many examples that have been cited in the past, such as the Bora and Kayapo of the Amazon (Altieri and Hecht, 1990), the Miskito of Nicaragua and Honduras (Neitchman, 1973), and the Maya of Mexico (Gomez-Pompa, et al., 1987) do not really farm inside a rain forest. They cut a piece of rain forest down in order to do slash and burn agriculture. While it is true, as we argue in Chapter 5, that this form of agriculture is really not very destructive, it nevertheless is agriculture in an area in which the trees of the tropical forest have been cut down.

[11] These ideas are fully explored in Chapters 4 and 5.

[12] Huston (1994) has made this point from a strictly empirical point of view, focusing on the preservation of biodiversity per se.

[13] This idea, which we suspect is far more common than the literature would suggest, has been explored in considerable detail by Chacon and Gliessman (1982) in the context of the *buen monte* (good weeds) versus *mal monte* (bad weeds) system of the lowland Maya in Mexico.

[14] Brownrigg, 1985; Soemarrvato, et al., 1985.

[15] Personal observations and Michon, 1983.

[16] Stoler, 1978.

[17] Redclift, 1987.

[18] Wilkin, 1987.

[19] Chapin (1988) has taken to task those who would propose such models as tried and true technology. While it is true that the Chinampa system, for example, is an elegant theoretical system, much research remains to be done before it can be regarded as a viable alternative.

Chapter 4

[1] This point is discussed in detail in Chapter 6. Also see Westoby, 1989.

[2] This analysis is based on the pioneering work of Boserup, 1965. The specific presentation here is after Minc and Vandermeer, 1990.

[3] In the Americas, potential draft animals, such as horses or oxen, had gone extinct before agriculture developed a role for them.

[4] McCann, 1976.

[5] Ibid.

[6] A great deal has been written on the role of the United Fruit Company in Guatemala. Key references include Melville and Melville, 1971; Immerman, 1982; and Rabe, 1988.

[7] This is a crucial conceptual issue that will be discussed in detail in the next chapter.

[8] Quoted in the film *Controlling Interest.*

[9] Rabe, 1988.

[10] There has been a great deal of speculation as to why the CIA was dispatched to overthrow a democratically elected government. The official line is that the Arbenz government was a Communist one, or at least very sympathetic to Communism, and therefore could not be allowed to prosper. In the hysteria of the times, this could very well have been a genuine reason. But any sensible analysis of the Arbenz program would have easily demonstrated that its program was far from what Moscow might have wanted. A more cynical interpretation suggests that the Dulles brothers, with their important connections to the Sullivan and Cromwell law firm, were simply protecting their own investments in the United Fruit Company. This is the impression one gets from McCann (1976). The truth is probably a little bit of both.

[11] This is the imagery first provided by Lewontin, 1982.

[12] Schob, 1975.

[13] In his famous and remarkably insightful analysis of the general tendencies of modern agriculture, Carey McWilliams warned of the immense problems we faced because of the tendencies of modern agriculture. His *Factories in the Fields* (1939) was the first convincing critique of modern agricultural production, calling for its transformation well before the development of a serious alternative agriculture movement.

[14] The purposeful development of this metaphor as a marketing tool is described in detail by Russell, 2001.

[15] Rosset, et al., 2000; Lappé, et al., 1998.

[16] Lappé, et al., 1998.

[17] Lappé, et al., 1998.

[18] Morgan, 1979.

[19] Since Chevron purchased Gulf in 1984, the famous seven sisters became six—Chevron, British Petroleum, Royal Dutch/Shell, Exxon, Mobil, and Texaco.

Chapter 5

[1] Wallerstein, 1980.

[2] This section follows closely the analysis of Lewontin, 1982.

[3] This, of course, is a topic for a textbook. We do not pretend to summarize economics as an academic discipline, but rather hope to describe those components important in understanding how the Global South functions. Since our goal is to comprehend the loss of tropical rain forests, and almost all of them are located in the Global South, we need to understand issues of the Global South. Part of grasping such issues is understanding how the developed world impacts upon them. That is what this section seeks to articulate. The basic framework has been described many times before, classically through Ricardo or Marx, or more recently in standard textbooks (e.g., Samuelson, 1970).

[4] This is what is usually referred to as an underconsumption crisis (Haberler, 1958).

[5] For different perspectives, see Kanth, 1994.

[6] de Janvry, 1982.

[7] This was in the late 1970s, before the spread of Solidarismo, an anti-union movement that has more recently clamped a strong hold on labor organizing in Costa Rica.

[8] For example, shifting cultivation (slash and burn agriculture) practices in Africa account for 70 percent of the clearing of closed-canopy forest (Brown and Thomas, 1990). Estimates of the number of farmers engaged in clearing forest lands in the humid tropics (including primary and secondary forests) each year range from 300 million to 500 million. Assessments of the area cleared by peasant farmers range from 7 million to 20 million hectares each year (National Research Council, 1993).

Chapter 6

[1] Westoby, 1989; Tucker, 2000.

[2] To a great extent this depends on the scale of the operation. If the area clear cut is extremely large, seed sources for the successional process may be unavailable.

[3] Some ecologists think that the actual number of species in an ecosystem increases as ecological succession proceeds, but only to a point. After that critical point, the diversity actually decreases, leading to the conclusion that a very old forest may be less diverse than a younger one.

[4] There is a problem with the definition here. Most ecologists today eschew the notion of a "mature" forest, and simply speak of "old growth." The notion of maturity implies something about a directed developmental sequence that does not fit well with what we now know about tropical rain forest succession.

[5] There are exceptions to this rule. Many swampy forests are characterized by the presence of only a few species. The biggest exception are the Southeast Asian dipterocarp forests, where the vast majority of trees in the forest belong to a single plant family, characterized by very large and straight trunks, a logger's delight.

[6] Johns, 1985.

[7] Ibid; Johns, 1991; Karr, 1971; Michael and Thornburgh, 1971; Webb, et al., 1977.

[8] On the other hand, there is much recent research concerning this topic (e.g., Butterfield, 1990; Espinoza and Butterfield, 1989).

[9] The anticipated benefit of such plantations is that they will provide much greater wood yields, thus taking pressure for harvesting off of the natural forests.

[10] More recently the practice has apparently been halted, and *Acacia* plantations may be a thing of the past in this area (J. Anderson, personal communication).

[11] Lohmann, 1990; Lugo, 1992.

[12] The organism is called the meliacian stem borer (*Hypsipila* sp.) and attacks any plant species in the family Meliaceae. Its effect is to turn the tree into a small bush, since the insect bores into the growing tip, causing the tree to sprout several other growing tips at the base of the damaged one. This process continues until you have a mahogany bush instead of a mahogany tree.

[13] Interview with representative of Ston Forestal published in the Costa Rican magazine *Panorama del Sur,* May 1993.

[14] There are, on the other hand, several studies that have attempted to accelerate the process of regeneration of rain forests after logging (e.g., Huss and Sutisna, 1991; Maury-Lechon, 1993).

[15] Blaikie and Brookfield, 1987.

[16] Mooney and Gordon, 1983; Pickett and White, 1985; Sousa, 1984. This point was covered with regard to tropical rain forests in Chapter 2.

[17] Pickett and White, 1985; and personal communication.

[18] The Palcazu project was designed according to the theoretical constructs of forest ecologist Hartshorn. Strips of forest were harvested in a clear-cut form, attempting to mimic the conditions normally found in forest light gaps (Hartshorn, 1987).

[19] Perfecto, et al., 1994; Boucher, et al., 1994; Yih, et al., 1991.

[20] Poore, 1989.

[21] Westoby, 1989.

[22] See Chapter 7.

Chapter 7

[1] Stiglitz, 2003.
[2] Johnson, 2004.
[3] Frank, 1998.

Chapter 8

[1] The history of the 1980s in Central America has been covered in many volumes. For example, Rosset and Vandermeer (1986), Barry (1987), and Edelmen and Kenen (1989), are excellent sources.
[2] Boza, 1993; Tangley, 1990.
[3] For a critique of these forestry laws see Thrupp, 1990.
[4] Seligson, 1980.
[5] World Resources Institute, 1990.
[6] Chico Mendez was the militant organizer of the rubber tappers in Brazil. His death at the hands of thugs hired by the cattle ranchers galvanized much of the rain forest preservation movement. In addition to his well-known activities on behalf of the rubber tappers, Mendez was a leader in Brazil's Workers Party, and an advocate of socialism as the only ultimate solution to the problems of environmental deterioration and social injustice (Mendez, 1989).
[7] A similar point of view, associated with African wildlife, was presented by Bonner, 1993.
[8] Children's author Dr. Seuss's mythical creature who spoke out against the ruthless exploitation of the trufula trees.
[9] This and the rest of the story of El Progreso comes from interviews with several residents of the community conducted in the summer of 1994.
[10] *Tico Times,* July 9, 1993.
[11] Document prepared by the Sarapiquí Association for Forests and Wildlife (undated).
[12] There is some debate as to how strong the influence of foreigners is in the association (personal interviews with local residents, 1994).
[13] Most of this section is taken from Vandermeer, 1991, and Perfecto, et al., 1994.
[14] In 1990 the FAO reported the following figures for rain forest extent in Central America (in 1,000 hectares): Costa Rica, 625; El Salvador, 33; Guatemala, 2,542; Honduras, 1,286; Nicaragua, 3,712; Panama, 1,802 (World Resources Institute, 1994).
[15] In 1986, Haiti's per capita GNP stood at $330, while that of Nicaragua was $790. For comparison, the Dominican Republic was $710 and Honduras was $740. All other Central American and Caribbean countries had GNPs larger than Nicaragua's. By 1991, Nicaragua's

GNP stood at $283, the worst in the hemisphere (Haiti stood at $375 at that point). We do not have more recent figures in Haiti, but subsequent to the ouster of Aristide and the economic blockade, we strongly suspect Haiti has returned to its unfortunate position as the poorest country in the hemisphere. Nicaragua is a close second.

[16] Vandermeer, et al., 1991.

[17] This was a widespread claim in both the Atlantic coast and Managua. Obviously its strict verification is hardly possible under current political circumstances.

[18] Martinez, 2004.

[19] Wright and Wolford, 2003.

[20] Kaimowitz, 1986.

[21] Current estimates suggest that about 18 percent of Cuba is covered in forest, which clearly represents an increase from the figures of 1959 (Rosset and Benjamin, 1994). Some of that area is clearly under plantations of trees and thus does not represent true tropical forest, but we are unable to determine how much. The FAO reported an annual deforestation rate in Cuba of 0.9 percent during the decade of the 1980s. We are, nevertheless, convinced that the forest cover has indeed increased since 1959. In the case of Puerto Rico, the island was almost completely deforested by the 1930s, where 99 percent of its primary forests were gone by that time and with an estimated cover of secondary forest of 10 to 15 percent (Lugo, 1988). It had increased its forested areas to 31 percent (284,000 hectares) by the late 1970s (Birdsey and Weaver, 1982).

[22] Unfortunately terms such as *sustainable development, ecological development, ecodevelopment*, and a variety of others have now been adopted and co-opted by the very agencies that have been promoting ecologically damaging development in the past, changing only the name of what they do.

Chapter 9

[1] As part of the normal process of photosynthesis, trees absorb carbon dioxide. If they are cut down, all of the carbon contained within them will be released into the atmosphere, thus contributing to greenhouse gases and global warming.

[2] It is estimated that approximately 80 percent of all insect species, 60 percent of all plant species, and 90 percent of all primate species live in rain forests (Westoby, 1989).

[3] Johns, 1991; Karr, 1971; Michael and Thornburgh, 1971; Lovejoy, et al., 1984; Lovejoy, et al., 1986.

[4] Swift, et al., (1994) note that as agroecosystems are transformed the pat-

tern of biodiversity may take many different forms, and we are largely ignorant which of these forms actually occurs in nature.

[5] Pimentel, et al., 1992.

[6] This includes protected categories such as forest reserves, wildlife preserves, protection zones, biological reserves, national parks, and recreational areas.

[7] Swift, et al., op cit.

[8] For example, several studies suggest that insect biodiversity is larger in systems where more than one crop are grown at the same time, that is, where planned biodiversity is higher than in the normal modern monoculture but, of course, much less than in the natural ecosystem that it replaced (Risch, et al., 1983; Andow, 1991). Furthermore, considering the organisms that live in the soil, their biodiversity pattern may very well show a type II response (Swift, et al., 1979).

[9] Gliessman, 1993; Dover and Talbot, 1987; Altieri, et al., 1987; Paoletti, 1988.

[10] Structural adjustment programs designed by the World Bank and IMF are likely to accelerate this process of agricultural intensification. In addition, the FAO's global agricultural strategy for the future focuses on agricultural intensification.

[11] Bonner's study of the conservation community concerned with the African elephant suggests that the drive for ever greater contributions may actually cause this sort of concern to even overlook the fact that the species of concern is harmed by fund-raising tactics (Bonner, 1993).

[12] Arthropods include insects, spiders, crabs, lobsters, and other such creatures with jointed legs.

[13] Ferraz, et al., 2003; Newmark, 1995.

[14] Vandermeer, et al., 2005.

[15] The data on arthropod diversity in coffee ecosystems is from our studies in Costa Rica (Perfecto and Snelling, 1995; Perfecto and Vandermeer, 1994).

[16] See note 13. These studies are cooperative ventures between the National Autonomous University of Costa Rica and the University of Michigan.

[17] Erwin and Scott, 1980; Wilson, 1987; Adis, et al., 1984.

[18] Perfecto and Vandermeer, 2002.

[19] See, for example, Shiva, 1993.

[20] Angelsen and Kaimowitz, 2001.

[21] Significantly complicating this picture is the freeze that hit Brazil in 1994. Coffee prices soared as news of coffee plantations being destroyed by the unexpected cold wave in Brazil reached the world.

[22] Due to overproduction, the year 2002 saw the lowest coffee prices paid

to producers in 30 years.
[23] See Shiva, 1993.

Chapter 10

[1] Vandermeer, 1995.
[2] Rocha, 1905.
[3] Meyers, 1985.
[4] The similarity between the terminology here and modern literary criticism is not accidental. Much of our analysis was at least partially stimulated by Eagleton (1983).
[5] Brechen, et al., 2000.

REFERENCES

Adis, J., Y. D. Lubin, and G. G. Montgomery. 1984. Arthropods from the canopy of inundated and Terra firme forests near Manaus, Brazil, with critical considerations on the pyrethrum-fogging technique. *Studies on Neotropical Fauna and Environment* 19:223–236.

Altieri, M. A., M. K. Anderson, and L. C. Merrick. 1987. Peasant agriculture and the conservation of crop and wild plant resources. *Conservation Biology* 1:49–58.

Altieri, M. A., and S. B. Hecht. 1990. *Agroecology and Small Farm Development.* New York: Macmillan.

Andow, D. A. 1991. Vegetational diversity and arthropod population response. *Annual Review of Entomology* 36:561–586.

Angelsen, A., and D. Kaimowitz (eds.). 2001. *Agricultural Technologies and Tropical Deforestation.* New York: CABI Publishing.

Barry, T. 1987. *Roots of Rebellion: Land and Hunger in Central America.* Boston: South End Press.

Birdsey, R. A., and P. L.Weaver. 1982. The forest resources of Puerto Rico. USDA Forest Service. Southern Forest Experiment Station. *Resource Bulletin* SO–85, October 1982.

Blaikie, P., and H. Brookfield (eds.). 1987. *Land Degradation and Society.* London: Methuen.

Bonner, R. 1993. *At the Hands of Man: Perils and Hope for Africa's Wildlife.* New York: Alfred A. Knopf.

Boserup, E. 1965. *The Conditions of Agricultural Growth.* Chicago: Aldine.

Boucher, D., J. H. Vandermeer, M. A. Mallona, N. Zamora, and I. Perfecto. 1994. Resistance and resilience in a directly regenerating rainforest: Nicaraguan trees of the Vochysiaceae after hurricane Joan. *Journal of Forest Ecology* 68:127–136.

Boza, M. A. 1993. Conservation in action: Past, present and future of the National Park System of Costa Rica. *Conservation Biology* 7:239–247.

Brechin, S., C. Fortwangler, P. Wilshusen, and P. C. West. 2000. The backlash to people-sensitive conservation and the future of international biodiversity conservation management. *Society and Natural Resources.*

Brown, H. C. P., and V. G. Thomas. 1990. Ecological considerations for the future of food security in Africa. In R. Edwards, R. Lal, P. Madden, R. H. Miller, and G. House, (eds.), *Sustainable Agricultural Systems.* Ankeny, IA: Soil and Water Conservation Society.

Brownrigg, L. A. 1985. *Home Gardening in International Development. What the Literature Shows.* The League for International Food

Education. Washington, DC: U.S. Agency for International Development.
Butterfield, R. 1990. Native species for reforestation and land restoration: A case study from Costa Rica. *Proceedings IUFRO World Congress*, Montreal. 3–14.
Butterfield, R. 1994. The regional context: Land colonization and conservation in Sarapiquí. In L. A. McDade, K. S. Bawa, H. A. Hespenheide, and G. S. Hartshorn (eds.), *La Selva: Ecology and Natural History of a Neotropical Rain Forest*. Chicago: University of Chicago Press.
Chacon, J. C., and S. R. Gliessman. 1982. Use of the "non-weed" concept in traditional tropical agroecosystems of southeastern Mexico. *Agroecosystems* 8:1–11.
Chapin, M. 1988. The seduction of models: Chinampa agriculture in Mexico. *Grassroots Development* 12:8–17.
de Janvry, A. 1982. *The Agrarian Question and Reformism in Latin America*. Baltimore: John Hopkins University Press.
Denslow, J. S. 1987. Tropical rainforest gaps and tree species diversity. *Annual Review of Ecology and Systematics* 18:431–451.
Dover, M., and L. Talbot. 1987. *To Feed the Earth: Agroecology for Sustainable Development*. Washington, DC: World Resources Institute.
Edelman, M., and J. Kenen (eds.). 1989. *The Costa Rican Reader*. New York: Grove.
Erwin, T. L., and J. C. Scott. 1980. Seasonal and size patterns, trophic structure, and richness of Coleoptera in the tropical arboreal ecosystem: The fauna of the tree *Luehea seemannii* Triana and Planch in the canal zone of Panama. *The Coleopterists Bulletin* 34:305–322.
Espinoza, M., and R. Butterfield. 1989. Adaptabilidad de 13 especies nativas maderables bajo condiciones de plantación en las tierras bajas húmedas del Atlántico de Costa Rica. In R. Salazar (ed.), *Manejo y aprovechamientode plantaciones forestales con especies de uso múltiple*. Guatemala: Actas Reunión IUFRO; Turrialba, Costa Rica: CATIE. 159–172.
Ferraz, G., G. J. Russell, P. C. Stouffer, R. O. Bierregaard, Jr., S. L. Pimm, and T. E. Lovejoy. 2003. Rates of species loss from Amazonian forest fragments. *Proceedings of the National Academy of Sciences* 100:14069–14073.
Frank, A. G. 1998. *Reorient: Global Economy in the Asian Age*. Berkeley: University of California Press.
Gliessman, S. R. 1993. Managing diversity in traditional agroecosystems of tropical Mexico. In C. S. Potter, J. I. Cohen, and D. Janczewski (eds.), *Perspectives on Biodiversity: Case Studies on Genetic Resource*

Conservation and Development. Washington, DC: American Association for the Advancement of Science Press. 65–74.

Gómez-Poma, A., J. S. Flores, and V. Sosa. 1987. The "Pet Kot": A man-made tropical forest of the Maya. *Interciencia* 12:10–15.

Haberler, G. 1958. *Prosperity and Depression.* Cambridge, MA: Harvard University Press.

Hardin, G. 1968. The tragedy of the commons. *Science* 162:1243–1248.

Hartshorn, G. S. 1987. Application of gap theory to tropical forest management: Natural regeneration on strip clear-cuts in the Peruvian Amazon. *Ecology* 70:567–569.

Hölldobler, B., and E. O. Wilson. 1990. *The Ants.* Cambridge, MA: Harvard University Press.

Huss, J., and M. Sutisna. 1991. Conversion of exploited natural Dipterocarp forests into semi-natural production forests. In H. Lieth and M. Lohmann (eds.), *Restoration of Tropical Forest Ecosystems.* Dordrecht, Netherlands: Kluwer Academic Publishers. 145–153.

Huston, M. 1993. Biological diversity. Soils and economics. *Science* 262:1676–1680.

Huston, M. 1995. *Biological Diversity: The Coexistence of Species in Changing Landscapes.* Cambridge: Cambridge University Press.

Immerman, R. H. 1982. *The CIA in Guatemala: The Foreign Policy of Intervention.* Austin: University of Texas Press.

Janzen, D. H., and T. W. Schoener. 1968. Differences in insect abundance and diversity between wetter and drier sites during a tropical dry season. *Ecology* 49:96–110.

Johns, A. D. 1985. Selective logging and wildlife conservation in tropical rain forest: Problems and recommendations. *Biological Conservation* 31:355–375.

Johns, A. D. 1991. Responses of Amazonian rain forest birds to habitat modification. *Journal of Tropical Ecology* 76:417–437.

Johnson, C. 2004. *Blowback: The Costs and Consequences of American Empire.* New York: Owl Books.

Kaimowitz, D. 1986. Nicaragua's Agrarian Reform: Six Years Later. In P. Rosset and J. Vandermeer (eds.), *The Nicaraguan Reader: Documents of a Revolution Under Fire.* New York: Grove.

Kanth, R. (ed.). 1994. *Paradigms in Economic Development.* Armonk, NY: M.E. Sharpe.

Karr, J. R. 1971. Structure of avian communities in selected Panama and Illinois habitats. *Ecological Monographs* 41:207–233.

Lappé, F. M., J. Collins, and P. Rosset. *World Hunger: Twelve Myths.* New York: Grove.

Lewis, S. A. 1992. Banana bonanza: Multinational fruit companies in Costa Rica. *The Ecologist* 22:289–290.

Lewontin, R. 1982. Agricultural research and the penetration of capital. *Science for the People* 14:12–17.

Lohmann, L. 1990. Commercial tree plantations in Thailand: Deforestation by any other name. *The Ecologist* 20(1):9–17.

Lovejoy, T. E., R. D. Bierregaard, Jr., A. B. Rylands, J. R. Malcolm, C. F. Quintela, L. H. Harper, K.S. Brown, Jr., A. H. Powell, G. V. N. Powell, H. O. R. Schubart, and M. B. Hays. 1986. Edges and other effects of isolation on Amazon forest fragments. In M. E. Solé (ed.), *Conservation Biology: The Science of Scarcity and Diversity.* Sunderland, MA: Sinauer. 257–285.

Lovejoy, T. E., J. M. Rankin, R. D. Bierregaard, Jr., K. S. Brown, Jr., L. H. Emmons, and M. E. Van der Voort. 1984. Ecosystem decay in Amazon forest remnants. In M. H. Niteki (ed.), *Extinctions.* Chicago: University of Chicago Press. 295–321.

Lugo, A. 1988. Estimating reductions in the diversity of tropical forest species. In E. O. Wilson (ed.), *Biodiversity.* Washington, DC: National Academy Press.

Lugo, A. 1992. Comparison of tropical tree plantations with secondary forest of similar age. *Ecological Monographs* 62(1):1–41.

Mabberley, D. J. 1992. *Tropical Rain Forest Ecology.* New York: Chapman and Hall.

Martinez, M. 2004. Antes invasion campesino: Miskitos defienden tierras. Managua: *La Prensa,* February 8.

Maury-Lechon, G. 1993. Biological characters and plasticity of juvenile tree stages to restore degraded tropical forests. In H. Lieth and M. Lohmann (eds.), *Restoration of Tropical Forest Ecosystems.* Dordrecht, Netherlands: Kluwer Academic Publishers. 37–46.

McCann, T. P. 1976. *An American Company: The Tragedy of United Fruit.* New York: Crown.

McWilliams, C. 1935. *Factories in the Field.* Berkeley: University of California Press (reprint edition published 2000).

Melville, T., and M. Melville. 1971. *Guatemala: The Politics of Land Ownership.* New York: Free Press.

Méndez, C. 1989. *Fight for the Forest: Chico Méndez in His Own Words.* London: Latin America Bureau.

Michael, E. D., and P. I. Thornburgh. 1971. Immediate effects of hardwood removal and prescribed burning on bird populations. *Southwest Naturalist* 15:359–370.

Michon, G. 1983. Village-forest gardens in West Java. In P. A. Huxley (ed.), *Plant Research and Agroforestry.* Nairobi: International Council for Research in Agroforestry (ICRAF). 13–24.

Minc, L. D., and J. H. Vandermeer. 1990. The origin and spread of agriculture. In C. R. Carroll, J. H. Vandermeer, and P. M. Rosset,

Agroecology. New York: McGraw Hill. 65–145.
Montangnini, F. 1994. Agricultural systems in the La Selva region. In L. A. McDade, K. S. Bawa, H. A. Hespenheide, and G. S. Hartshorn (eds.), *La Selva: Ecology and Natural History of a Neotropical Rain Forest.* Chicago: University of Chicago Press. 307–316.
Mooney, H. A., and M. Gordon (eds.). 1983. *Disturbance and Ecosystems.* Berlin: Springer-Verlag.
Morgan, D. 1979. *Merchants of Grain.* New York: Viking.
Myers, N. 1980. *The Conservation of Tropical Moist Forests.* Washington, DC: National Academy of Sciences.
Myers, N. 1985. *The Primary Source: Tropical Forests and Our Future.* New York: W.W. Norton and Co.
National Research Council. 1993. *Sustainable Agriculture and the Environment in the Humid Tropics.* Washington, DC: National Academy Press.
Newmark, W. D. 1995. Extinction of mammal populations in western North American national parks. *Conservation Biology* 9:512–526.
Nietchmann, B. 1973. *Between Land and Water: The Subsistence Ecology of the Miskito Indians, Eastern Nicaragua.* New York: Seminar Press.
Paoletti, M. G. 1988. Soil invertebrates in cultivated and uncultivated soils in northern Italy. *Firenze* 71:501–563.
Perfecto, I., M. A. Mallona, I. Granzow de la Cerda, and J. H. Vandermeer. 1994. Los recursos terrestres de la Costa Caribeña de Nicaragua: Hacia una filosofia de sostenibilidad. *Wani* 15:46–59.
Perfecto, I., and R. Snelling. 1995. Ant diversity in the coffee agroecosystem in Costa Rica. *Ecological Applications.*
Perfecto, I., and J. H. Vandermeer. 1994. The ant fauna of a transforming agroecosystem in Central America. *Trends in Agricultural Science* 2:7–13.
Perfecto, I., and J. H. Vandermeer. 2001. The quality of the agroecological matrix in a tropical montane landscape: Ants in coffee plantations in southern Mexico. *Conservation Biology* 16:174–182.
Phillips, O. L., and A. H. Gentry. 1994. Increasing turnover through time in tropical forests. *Science* 263:954–958.
Pickett, S. T. A., and P. S. White. 1985. *The Ecology of Natural Disturbance and Patch Dynamics.* New York: Academic Press.
Pimentel, D., V. Stachow, D. A. Takacs, H. W. Brubaker, A. R. Dumas, J. H. Meaney, J. A. S. O'Neal, D. E. Onsi, and D. B. Corzinus. 1992. Conserving biological diversity in agricultural/forestry systems. *Bioscience* 42:354–362.
Poore, D. 1989. *No Timber without Trees.* London: Earthscan Publication Ltd.
Rabe, S. G. 1988. *Eisenhower and Latin America: The Foreign Policy of*

Anticommunism. Chapel Hill: University of North Carolina Press.
Redclift, M. 1987. "Raised bed" agriculture in pre-Columbian Central and South America: A traditional solution to the problem of "sustainable" farming systems. *Biological Agriculture and Horticulture: An International Journal* 5:51–59.
Richter, D. D., and L. I. Babbar. 1991. Soil diversity in the tropics. *Advances in Ecological Research* 321:315–389.
Ricklef, R. E., and D. Schluter. 1993. *Species Diversity in Ecological Communities: Historical and Geographical Perspectives.* Chicago: University of Chicago Press.
Risch, S. J., D. Andow, and M. Altieri. 1983. Agroecosystem diversity and pest control: Data, tentative conclusions, and new research directions. *Environmental Entomology* 12:625–629.
Rocha, J. 1905. *Memorandum de un viaje.* Bogota: Editorial El Mercurio.
Rosset, P., and M. Benjamin. 1994. *The Greening of Cuba.* Melbourne: Ocean Press.
Rosset, P., J. Collins, and F. M. Lappé. 2000. Lessons from the Green Revolution. *Tikkun Magazine* March/April.
Rosset, P., and J. H. Vandermeer (eds.). 1986. *Nicaragua, Unfinished Revolution: The New Nicaraguan Reader.* New York: Grove Press.
Russell, E. 2001. *War and Nature: Fighting humans and insects with Chemicals from World War I to Silent Spring (Studies in Environment and History).* Cambridge : Cambridge University Press.
Samuelson, P. A. 1970. *Economics.* New York: McGraw-Hill.
Sanchez, P. A. 1976. *Properties and Management of Soils in the Tropics.* New York: Wiley.
Schob, D. E. 1975. *Hired Hands and Plowboys: Farm Labor in the Midwest.* Champaign: University of Illinois Press.
Schoener, T. W. 1993. On the relative importances of direct versus indirect effects in ecological communities. In H. Kawanabe, J. E. Cohen, and K. Iwasaki (eds.), *Mutualism and Community Organization: Behavioral, Theoretical and Food Web Approaches.* Oxford: Oxford University Press. 365–411.
Seligson, M. A. 1980. *Peasants of Costa Rica and the Development of Agrarian Capitalism.* Madison: University of Wisconsin Press.
Shiva, V. 1993. *Monocultures of the Mind.* London: Zed Books.
Soemarwoto, O., I. Soemarwoto, Karyano, E. M. Soekartadiredja, and A. Famlan. 1985. The Javanese home gardens as an integrated ecosystem. *Food Nutrition Bulletin* 7:44–47.
Sousa, W. P. 1984. The role of disturbance in natural communities. *Annual Review of Ecology and Systematics* 15:353–391.
Stiglitz, J. 2003. *Globalization and Its Discontents.* New York: W.W. Norton.

Stoler, A. L. 1978. Garden use and household economy in rural Java. *Bulletin of Indonesian Economic Studies* 14:85–101.

Swift, M. J., O. W. Heal, and J. M. Anderson. 1979. Decomposition in terrestrial ecosystems. *Studies in Ecology*, vol. 5. Oxford: Blackwell Scientific.

Swift, M., J. H. Vandermeer, J. Anderson, R. Ramakrishnan, B. Hastings, and C. Ong. 1994. Biodiversity changes in agricultural transformation. In H. Mooney (ed.), *Global Change and Biodiversity*.

Tangley, L. 1990. Cataloging Costa Rica's Diversity. *Bioscience* 40:633–636.

Terborgh, J. 1992. *Diversity and the Tropical Rain Forest*. New York: Scientific American Library.

Thrupp, L. A. 1990. Environmental Initiatives in Costa Rica: A political ecology perspective. *Society and Natural Resources* 33:243–256.

Thrupp, L. A. 1991. The human guinea pigs of Rio Frio: Standard Fruit keeps its eye on the bottom line. *The Progressive*, April: 28–30.

Tucker, R. P. 2000. *Insatiable Appetite: The United States and the Ecological Degradation of the Tropical World*. Berkeley: University of California Press.

USDA (United States Department of Agriculture). 1975. *Soil Taxonomy. Agriculture Handbook #436*. Washington, DC: US Government Printing Office.

Vandermeer, J. H. 1991. The political economy of sustainable development: The Southern Atlantic Coast of Nicaragua. *The Centennial Review* XXXV:265–294.

Vandermeer, J. H. 1995. *Reconstructing Biology: Genetics and Ecology in the New World Order*. New York: Wiley.

Vandermeer, J. H., D. Boucher, and I. Perfecto. 1991. Conservation in Nicaragua and Costa Rica: Indirect consequences of social policy. *INTECOL Bulletin*, January.

Vandermeer, J. H., M. A. Mallona, D. Boucher, K. Yih, and I. Perfecto. 1995. Three years of ingrowth following catastrophic hurricane damage on the Caribbean Coast of Nicaragua: Evidence in support of the direct regeneration hypothesis. *Journal of Tropical Ecology* 11:465–471.

Vandermeer, J. H., I. Perfecto, S. Philpott, and M. J. Chappell. 2005. *Refocusing Conservation in the Landscape: The Matrix Matters*. Turrialba, Costa Rica: CATIE.

Wallerstein, I. 1980. *The Modern World System II: Mercantilism and the Consolidation of the European World-Economy, 1260–1750*. New York: Academic Press.

Webb, W. L., D. F. Behrend, and B. Saisoin. 1977. Effect of logging on songbird populations in a northern hardwood forest. *Wildlife*

Monographs 55.
Werner, E. E. 1992. Individual behavior and higher-order species interactions. *American Naturalist* 140:S5–S32.
Westoby, J. 1989. *Introduction to World Forestry.* Oxford: Basil Blackwell.
Whitmore, T. C. 1991. *An Introduction to Tropical Rain Forests.* Oxford: Clarendon Press.
Wilkin, G. 1987. *Good Farmers.* Berkeley: University of California Press.
Wilson, E. O. 1987. The arboreal ant fauna of Peruvian Amazon forests: A first assessment. *Biotropica* 19:245–251.
Wilson, E. O. 1988. The current state of biological diversity. In E. O. Wilson (ed.), *Biodiversity.* Washington, DC: National Academy Press. 3–8.
World Resources Institute. 1990. *World Resources.* Oxford: Oxford University Press.
World Resources Institute. 1994. *World Resources.* Oxford: Oxford University Press.
Yih, K., D. Boucher, J. H. Vandermeer, and N. Zamora. 1991. Recovery of the rainforest of southeastern Nicaragua after destruction by Hurricane Joan. *Biotropica* 23:106–113.

INDEX

A

acacia, 25–26, 87, 185
acid soils, 37–38, 45, 181
AFL-CIO, 8
African rain forests, 33, 184
agrarian reform, 55–57, 117–18, 131–32
agriculture
- *See also* modern agriculture
- biodiversity and, 141–47, 152–59
- Chinampa system of, 47–48, 49, 182
- difficulties for, in rain forests, 35–37
- enclave production and, 53–60, 67
- export, 79–80
- future models for, 47–49
- origin and intensification of, 50–53
- with perennial vs. annual crops, 45–47
- recovery from, 92
- slash and burn, 41–43, 51, 184
- stabilizing, on rain forest soils, 43–47
- sustainable, 12
- in the United States, 70–71

Aleman, Arnoldo, 119–20
alluvial soils, 38, 45, 181
Almolonga Valley, 48–49
andisols, 181
André, 66
Anglo-Persian, 98, 104
annual crops, 47
ants
- acacia plants and, 25–26
- species of, 20

Arbenz, Jacobo, 55, 57, 98, 182–83
Archer Daniels Midland, 66
Armas, Castillo, 57
articulation, 76
ASBANA (Asociación Bananera Nacional), 9
Association for Free Labor Development, 8
Aztecs, 47

B

Baker, Lorenzo Dow, 54
banana plantations
- *See also* individual companies
- environmental consequences of, 3–4, 5–6
- expansion of, 5, 7–10
- history of, 6–7, 55–57
- layoffs from, 6, 60, 74, 126–27
- modern, 58–60
- research industries for, 9
- worker safety and, 5

Barva volcano, 1, 4
beef industry, 3–4
Bhopal, 78
biodiversity
- agriculture and, 141–47, 152–59
- causes of, 20–21
- credits, 159
- fragmented landscapes and, 147–49
- logging's effects on, 140–41
- models for preserving, 154–60
- planned vs. associated, 142–44, 153
- rain forest stability/fragility and, 17–18
- significance of, 20, 138–39
- small organisms and, 146–47
- utilitarianism and, 139–41

birds
- pollination by, 22, 24
- species of, 20
- watching, 139–40

birth control, 11
Bluefields, 115, 116, 117, 147, 148

Bolaños, Enrique, 120
Borneo, 86
Boston Fruit Company, 54, 55
Brazil, 128, 189
Bretton Woods conference, 95–97, 99–100
Bridges for Peace, 8
Britain, 53–54, 94, 97, 101, 139
British Petroleum, 104
Bunge, 66
Bush, George H., 99
Bush, George W., 99, 136

C

calcium, 29
capitalism, effects of, 11, 12
carbon dioxide, 28, 188
Cargill, 66, 70
Castle and Cook, 56
CEDEHCA (Center for Human, Civil, and Autonomous Rights), 125
China, 97
Chinampa system, 47–48, 49, 182
Chiquita Brands, 101–3, 112, 126, 128, 129, 130
Churchill, Winston, 98
CIA, 8, 57, 98, 99, 117, 182
CIDCA (Center for Research and Documentation of the Atlantic Coast), 125
Civil War, 62
clays, 30–31
clear-cutting, 82
climax species, 27
Clinton, Bill, 102
coffee, 126, 149–54, 158–59, 189
cold war, 57, 97
colonialism, 53–54, 68, 97, 101–2, 139
conservation
  in Costa Rica, 15, 106–7, 141–42, 168
  failures of programs for, 167–69
  of large vs. small organisms, 146–47
  mainstream approach to, 13, 134
  needs of local people and, 167– 68
  theoretical principle of, 165–66
  viewing, at landscape level, 147–49
Constantinople, 64, 65
Continental, 66
CORBANA (Corporación Bananera Nacional), 7, 9, 174, 175
corn, 45–46, 144
Costa Rica
  agrarian reform in, 131–32
  banana industry in, 5, 7–10, 125–30
  climate of, 1
  comparison between Nicaragua and, 131–32
  conservation ethic of, 15, 106–7, 141–42, 168
  debt of, 179
  deforestation in, 106, 107, 119, 125–26
  immigration to, from Nicaragua, 126, 129–30
  labor unions in, 8, 128
  railroad in, 54–55
  social security network in, 127
  tree plantations in, 88
  United Fruit Company in, 55, 56, 59, 98
cotton, 64
Creoles, 120, 121, 124
Cuba, 132, 133, 187

D

Danum Valley, 86–87
DBCP, 5
DDT, 62
debt, 179

deforestation
causes of, 11–13, 79–80, 130–31, 169–70, 171–74
in Costa Rica, 106, 107, 119, 125–26
in Cuba, 187
defining, 89, 147
models for ending, 13–14, 174–76
in Nicaragua, 117
in Puerto Rico, 187–88
in Sarapiquí, 6, 11
slash and burn agriculture as, 51
stages of, 91–92
Del Monte, 129
dependency theory, 73–76, 78
Dipterocarpaceae, 33, 185
disturbance ecology, 89–92
Dole, 6, 112, 126, 129. *See also* Standard Fruit Company
Dominican Republic, 187
Dow Chemical, 70
draft animals, 53, 182
dualism, 76–79
Dulles, Allen, 57, 183
Dulles, John Foster, 57, 183

E

ecological development, 13
ecological succession, 27, 41, 43, 81–82, 86, 184
economics
of the developed world, 71–72
dualism and, 76–79
of the Global South, 73–76
ecotourism, 111–12, 114
Ecuador, 129
Eisenhower, Dwight, 57
elephants, 146, 189
El Progreso, 109–10, 114, 125–28, 186
enclave production, 53–60, 67
enemies hypothesis, 22, 148
energy use, 104
English East India Company, 53–54
entisols, 181
environmental movement, mainstream, 13, 134
exchange rate stabilization, 95–96
export agriculture, 79–80
extinction, 146

F

fertilizers, 63
flowering, synchronous mass, 23–24
food insecurity, 12–13, 80
futures trading, 65, 66

G

Gandhi, 97
gap dynamics, 27–29
Garifuna, 116
GATT (General Agreement of Tariffs and Trade), 101
Geest, 112
Gerika, 109–10
globalization, 100–105, 136
global warming, 167, 188
gold standard, 99
grain trade, 64–67
Green Revolution, 63
Guatemala, 48–49, 55–57, 98, 100, 104
Guatemala City, 57
Gulf War, 99

H

Haiti, 187
hamburger connection, 3–4
herbivores, 19, 24–26, 36–37
hillside soils, 39, 181
histosols, 181
Holbrook Travel, 111–12
Honduras, 55, 56, 75, 98, 187
Hong Kong, 79

hummingbirds, 22, 24
humus, 30–31, 181

I

IDA (Institute of Agrarian Development), 131–32
inceptisols, 181
India, 53–54, 78, 97
insects
  biodiversity of, 188
  as herbivores, 24
  pollination by, 22, 24
  species of, 20
Intel, 126
International Monetary Fund (IMF), 14, 15, 94, 95, 96, 103, 189
Iran, 97–98, 100, 104
Iraq, invasion of, 94, 99, 136

J

Jamaica, 54
Japan, 94, 95
Java, 46
John Deere, 70

K

Keith, Minor C., 3, 54–55
Kota Kinambalu, 86
Kurinwas River, 121

L

labor unions, 8, 10, 75, 128
La Martita, 111, 114
light gaps, 19, 21, 27–28, 82, 91
Limón, 55
Lindner, Carl, 102, 103
Liverpool, 64, 65
logging
  biodiversity and, 140–41
  clear-cutting, 82, 92
  direct effects of, 81–82, 84–85, 91, 140–41
  history of, 6–7, 10
  in the RAAS, 118
  recovery from, 50, 81–82, 140–41
  reforestation and, 85–89
  secondary damage from, 82, 84, 91
  selective, 82, 84–85, 92
Loma de Mico, 115
Louis-Dreyfus, Leopold, 65–66, 67
Louis-Dreyfus Company, 66

M

magnesium, 29
mahogany, 88, 185
Malaysia, 86–87
Managua, 124
mast years, 27
Maya, 47, 48–49, 49, 182
McCormick, Cyrus, 61–62
megafauna, 146
meliacian stem borer, 185
melina, 161
Mendez, Chico, 107, 186
mestizos, 120, 122, 124
metapopulation dynamics, 169
Mexico, 48, 144, 153, 177
MI6, 98
milling, 64
Miskitos, 116, 120, 121–22, 124
modern agriculture
  components of, 69–71
  effects of, 50
  history of, 53, 61–67
  overproduction and, 128–29, 158–59
  popular conception of, 60–61
  recovery from, 92
  socioeconomic side of, 63–67
  technical side of, 61–63
  transfer of goods in, 70

transfer of money in, 70
tropical deforestation and, 79–80
modern world system
economics of, 71–79
emergence of, 68–69
interconnectedness of, 69
morning glory vines, 45–46
Mossadegh, Mohammed, 97–98
MST (Landless Workers' Movement), 128
mutualism, 22, 25–26

N

National Institute for Biodiversity (Costa Rica), 139
nature construction, 164–67
nature worship, 161
Nicaragua
agrarian reform in, 131
agricultural expansion in, 115
Caribbean coast tensions in, 119–25
comparison between Costa Rica and, 131–32
deforestation in, 117
distribution of rain forests in, 115
GNP of, 187
migration from, into Costa Rica, 126, 129–30
peoples of, 116
politics in, 110, 116, 117, 119–20, 121, 123
tree plantations in, 88
nitrogen, 29
Nixon, Richard, 57, 99–100
Nobel Group, 66
North American Free Trade Agreement (NAFTA), 94
nutrient cycling, 42–43

O

Oates, John, 167
Odessa, 64, 65, 66, 67
Organization for Tropical Studies, 110, 112
overpopulation, 11, 12, 63, 130, 169–70
overproduction, 128–29, 158–59, 189
oxisols, 181

P

Palcazu project, 89, 185–86
Patchy River, 118
peasants
agrarian reform and, 55, 117, 131–32
displacement of, from land, 55, 119
legal status of, 12
organizations of, 105
perennial crops, 45–47
Peru, 89
pesticides, 62–63, 78, 144
pharmaceutical companies, 139
pioneer species, 27, 86
plant defenses, 24–26
PLC (Constitutional Liberal Party), 119–20, 123–25
plows, 53
poisonous plants, 25
political action plan, 176–77
political ecology strategy, 13–14
pollination, 19, 22–24
population growth, 51–52. *See also* overpopulation
Porto Alegre, 105
potassium, 29
poverty, 12, 63, 97
Preston, Andrew, 54
Puerto Rico, 78, 132–33, 187–88
Puerto Viejo, 1, 4–5

R

RAAS (Región Autonoma del Atlántico Sur), 2, 115–18
rain forests
  biodiversity in, 3, 18, 20–22, 138, 188
  causes of destruction of, 11–13, 79–80, 130–31, 169–70, 171–74
  complexity of, 16–17, 34
  conservation of, 15, 106–8
  difficulties for agriculture in, 35–37
  disturbance ecology and, 89–92
  diversity of, 32–34
  environmental role of, 3
  functioning of, 18–32
  future of, 14
  gap dynamics in, 19, 27–29
  herbivores in, 19, 24–26
  location of, 2, 16, 17
  models for saving, 13–14, 135–37, 165–69, 174–76
  pattern of transformation of, 4
  pollination in, 19, 22–24
  recovery of, 29, 89–92
  seed dispersal in, 19, 26–27
  size of, 3
  social construction of, 161–65, 166
  social justice and, 169–70, 176–77
  soils of, 19, 32, 35–36, 37–41, 180–81
  stability vs. fragility of, 17–18, 21, 34
Ramas, 116
reapers, automatic, 61–62
reforestation, 85–89, 161
Rio Frio, 179
road building, 82, 84
rubber tappers, 186
Russia, 64–65

S

Sabah, 86–87
Sabah Forest Industries, 87
Sandinistas, 116, 117, 120, 123
San Jose, 55
San Juan River, 1
Sarapiquí
  banana expansion in, 5, 7–10, 112
  deforestation of, 6, 11
  infrastructure of, 8–9
  land use in, 5, 111
  location of, 1, 2, 109
  political groups in, 109–11
  possible program for, 113–15
  rain forest distribution in, 2, 6
  tourism in, 111–12, 114
  as useful example, 4
Sarapiquí Association for Forests and Wildlife, 111, 114
satiation, 27
Seattle, 94, 103–4, 105, 136
secondary forests, 82
seed dispersal, 19, 26–27
selfing, 23
Selous National Park, 147
Selva Verde Lodge, 111–12
Shah of Iran, 98
shifting cultivation. *See* slash and burn agriculture
Singapore, 79
slash and burn agriculture, 41–43, 51, 184
Smith Forestal, 88
social justice, 169–70, 176–77
soils
  acid, 37–38, 45, 181
  acidity gradient in, 29–30, 31, 32
  alluvial, 38, 45, 181
  clays in, 30–31
  fertility of, 29–32, 39–40
  heterogeneity of tropical, 180
  hillside, 39, 181
  iron in, 181

organic matter in, 30–31, 180–81
of rain forests, 19, 32, 35–36, 37–41, 180–81
swamp, 39, 45, 181
volcanic, 38–39, 45, 181
Solidarista movement, 8, 10, 179, 184
South American rain forests, 33
South East Asian rain forests, 33, 185
South Korea, 78
Soviet Union, 95, 102
Standard Fruit Company, 5–6, 7, 179
Ston Forestal, 88, 161, 163
storms, effects of, 28–29, 89–91
structural adjustment programs (SAPs), 96, 189
Sullivan and Cromwell, 57, 183
Sumos, 116
sustainable agriculture, 12
sustainable development, 13, 134–35
swamp soils, 39, 45, 181

T

Taiwan, 78
Tanzania, 147
Tasba Pauni, 121
tea, 53–54, 139
Terborgh, John, 167
traplining, 24
tree plantations, 87–88, 161
trees
death of, 19, 27
distribution of species of, 22
fast- vs. slow-growing, 28
number of species of, 21

U

ultisols, 181
underconsumption crisis, 184
Union Carbide, 78
United Fruit Company, 3, 7, 55–57, 59, 74, 98, 101, 182, 183
United States
agriculture in, 70–71
Army Corps of Engineers, 8
banana companies and, 8–9, 54, 56–57, 102–3
foreign incursions of, 56–57, 98–99, 100, 133
status of, after World War II, 94, 99
UNO (United Nicaraguan Opposition), 116

V

Vietnam War, 98–99, 100, 104
volcanic soils, 38–39, 45, 181

W

Wallerstein, Immanuel, 69
weathering, 38
web of causality, 4, 171–74, 176, 177
wetlands, 39
wheat, 64–65
wild banana, 22, 45–46
World Bank, 15, 94, 95, 96–97, 103, 189
World Resources Institute, 106, 107
World Social Forum, 105
World Trade Organization (WTO), 15, 94, 101–4
World War II, 94, 95, 99

Y

YATAMA, 116

FOOD FIRST BOOKS OF RELATED INTEREST

To Inherit the Earth:
The Landless Movement and the Struggle for a New Brazil
Angus Wright and Wendy Wolford
The story of one of the world's most successful contemporary grassroots movements, Brazil's Landless Workers Movement (MST). The authors put the movement in its historical, political, and environmental context, trace its growth and organization, sum up its accomplishments and setbacks, and analyze the issues the MST faces going forward.
Paperback, $15.95
ISBN: 0-935028-90-0

Benedita da Silva: An Afro-Brazilian Woman's Story of Politics and Love
As told to Medea Benjamin and Maisa Mendonça
Foreword by Jesse Jackson
The inspiring memoir of a woman who overcame poverty and tragedy to become one of the most prominent policians in Brazil. Benedita da Silva shares the story of her life as an advocate for the rights of women, people of color, and the poor, and argues persuasively for economic and social human rights in Brazil and everywhere.
Paperback, $15.95
ISBN: 0-935028-70-6

The Future in the Balance: Essays on Globalization and Resistance
Walden Bello
Edited with a preface by Anuradha Mittal
A new collection of essays by Third World activist and scholar Walden Bello on the myths of development as prescribed by the World Trade Organization and other institutions, and the possibility of another world based on fairness and justice.
Paperback, $13.95
ISBN: 0-935028-84-6

Views from the South: The Effects of Globalization and the WTO on Third World Countries
Foreword by Jerry Mander
Afterword by Anuradha Mittal
Edited by Sarah Anderson
This rare collection of essays by Third World activists and scholars describes in pointed detail the effects of the WTO and other Bretton Woods institutions.
Paperback, $12.95
ISBN: 0-935028-82-X

Call our distributor, CDS, at (800) 343-4499 to place book orders. All orders must be pre-paid.

ABOUT FOOD FIRST

Food First, also known as the Institute for Food and Development Policy, is a nonprofit research and education-for-action center working to expose the root causes of hunger in a world of plenty. It was founded in 1975 by Dr. Joseph Collins and Frances Moore Lappé, author of the best selling book *Diet for a Small Planet.* Food First research has revealed that hunger is created by concentrated economic and political power, not by scarcity. Resources and decision-making are in the hands of wealthy few, depriving the majority of land and jobs, and therefore of food.

Hailed by the *New York Times* as "one of the most established food think tanks in the country," Food First has grown to profoundly shape the debate about hunger and development. Through books, reports, videos, media appearances, and speaking engagements, Food First experts reveal the often hidden roots of hunger, and show how individuals can get involved in ending the problem. Food First inspires action by bringing to light the courageous efforts of people around the world who are creating faming and food systems that truly meet people's needs.

BECOME A MEMBER OF FOOD FIRST

Individual member contributions provide more than half of the funds for Food First's work. Because Food First is not tied to government, corporate, or university funding, we can speak with a strong, independent voice. The success of our program depends on dedicated volunteers and staff, as well as financial support from our activist donors. Your gift will help strengthen our effort to improve the lives of hungry people around the world.

I would like to become a Food First member!

Enclosed is my tax-deductible contribution of:

☐ $35 ☐ $50* ☐ $500*

☐ $40 ☐ $100* ☐ $1,000* ☐ Other: $______

All gifts are tax-deductible.

Method of Payment

☐ Check or money order enclosed. All foreign orders must be in US funds. Make checks out to Food First. Send to:

398 – 60th Street
Oakland, CA 94618
(510) 654-4400, FAX (510) 654-4551, www.foodfirst.org.

☐ Visa ☐ MC ☐ AmEx

Name on card ______________________________

Card number ____________________ Expiration date __________

Name ______________________________

Address ______________________________

____

City ____________________ State __________ Zip __________

Tel: (day) ____________________ Tel: (eve) ____________________

*A donation of $50 or more includes a FREE one-year subscription to the *New Internationalist.*